CLIMATE CHANGE EDUCATION

Engaging Family Private Forest Owners on Issues Related to Climate Change

A WORKSHOP SUMMARY

I08830Z8

Alexandra S. Beatty, Michael Feder, and Martin Storksdieck, *Rapporteurs*

Steering Committee on Engaging Family Private Forest Owners on Issues Related to Climate Change

Board on Science Education

Division of Behavioral and Social Sciences and Education

NATIONAL RESEARCH COUNCIL
OF THE NATIONAL ACADEMIES

THE NATIONAL ACADEMIES PRESS
Washington, D.C.
www.nap.edu

THE NATIONAL ACADEMIES PRESS 500 Fifth Street, NW Washington, DC 20001

NOTICE: The project that is the subject of this report was approved by the Governing Board of the National Research Council, whose members are drawn from the councils of the National Academy of Sciences, the National Academy of Engineering, and the Institute of Medicine. The members of the committee responsible for the report were chosen for their special competences and with regard for appropriate balance.

This study was supported by Award No. AG-3187-P-12-0040 between the National Academy of Sciences and the U.S. Department of Agriculture. Any opinions, findings, conclusions, or recommendations expressed in this publication are those of the author(s) and do not necessarily reflect the views of the organizations or agencies that provided support for the project.

International Standard Book Number-13: 978-0-309-30539-6
International Standard Book Number-10: 0-309-30539-X

Additional copies of this report are available from the National Academies Press, 500 Fifth Street, NW, Keck 360, Washington, DC 20001; (800) 624-6242 or (202) 334-3313; http://www.nap.edu.

Printed in the United States of America

Cover: All images are from iStock Photo, from the top—Close up of autumn leaf on oak bark; second from the top—Burnt hillside, environmental disaster; third from the top— Ranch in a forested valley in Oregon; bottom— Forester protecting trees from damage by deer.

Suggested citation: National Research Council. (2014). *Climate Change Education: Engaging Family Private Forest Owners on Issues Related to Climate Change, A Workshop Summary.* A. Beatty, M. Feder, and M. Storksdieck, *Rapporteurs.* Steering Committee on Engaging Family Private Forest Owners on Issues Related to Climate Change. Board on Science Education, Board on Environmental Change and Society, Division of Behavioral and Social Sciences and Education. Washington, DC: The National Academies Press.

THE NATIONAL ACADEMIES

Advisers to the Nation on Science, Engineering, and Medicine

The **National Academy of Sciences** is a private, nonprofit, self-perpetuating society of distinguished scholars engaged in scientific and engineering research, dedicated to the furtherance of science and technology and to their use for the general welfare. Upon the authority of the charter granted to it by the Congress in 1863, the Academy has a mandate that requires it to advise the federal government on scientific and technical matters. Dr. Ralph J. Cicerone is president of the National Academy of Sciences.

The **National Academy of Engineering** was established in 1964, under the charter of the National Academy of Sciences, as a parallel organization of outstanding engineers. It is autonomous in its administration and in the selection of its members, sharing with the National Academy of Sciences the responsibility for advising the federal government. The National Academy of Engineering also sponsors engineering programs aimed at meeting national needs, encourages education and research, and recognizes the superior achievements of engineers. Dr. C. D. Mote, Jr., is president of the National Academy of Engineering.

The **Institute of Medicine** was established in 1970 by the National Academy of Sciences to secure the services of eminent members of appropriate professions in the examination of policy matters pertaining to the health of the public. The Institute acts under the responsibility given to the National Academy of Sciences by its congressional charter to be an adviser to the federal government and, upon its own initiative, to identify issues of medical care, research, and education. Dr. Harvey V. Fineberg is president of the Institute of Medicine.

The **National Research Council** was organized by the National Academy of Sciences in 1916 to associate the broad community of science and technology with the Academy's purposes of furthering knowledge and advising the federal government. Functioning in accordance with general policies determined by the Academy, the Council has become the principal operating agency of both the National Academy of Sciences and the National Academy of Engineering in providing services to the government, the public, and the scientific and engineering communities. The Council is administered jointly by both Academies and the Institute of Medicine. Dr. Ralph J. Cicerone and Dr. C. D. Mote, Jr., are chair and vice chair, respectively, of the National Research Council.

www.national-academies.org

STEERING COMMITTEE ON ENGAGING FAMILY PRIVATE FOREST OWNERS ON ISSUES RELATED TO CLIMATE CHANGE

JAMES FINLEY (*Chair*), Forest Resources, The Pennsylvania State University
ARUN AGRAWAL, School of Natural Resources and Environment, University of Michigan, Ann Arbor
SHORNA ALLRED, Department of Natural Resources, Cornell University
CHARLES CANHAM, Cary Institute of Ecosystems
JOE E. HEIMLICH, Environment and Natural Resources, The Ohio State University
STEVEN KOEHN, Maryland Department of Natural Resources
MAUREEN H. MCDONOUGH, Department of Forestry, Michigan State University
MARY TYRRELL, School of Forestry and Environmental Studies, Yale University

MICHAEL A. FEDER, *Study Director*
MARTIN STORKSDIECK, *Director, Board on Science Education*
PAUL STERN, *Senior Scholar, Board on Environmental Change and Society*
REBECCA KRONE, *Senior Program Assistant*

BOARD ON SCIENCE EDUCATION

BOARD ON ENVIRONMENTAL CHANGE AND SOCIETY

Preface

This report has been prepared by the workshop rapporteurs as a factual summary of what occurred at the workshop. The planning committee's role was limited to planning and convening the workshop. The views contained in the report are those of individual workshop participants and do not necessarily represent the views of all workshop participants, the planning committee, or the National Research Council (NRC).

This workshop summary has been reviewed in draft form by individuals chosen for their diverse perspectives and technical expertise, in accordance with procedures approved by the Report Review Committee of NRC. The purpose of this independent review is to provide candid and critical comments that will assist the institution in making its published report as sound as possible and to ensure that the report meets institutional standards for objectivity, evidence, and responsiveness to the charge. The review comments and draft manuscript remain confidential to protect the integrity of the process. I would like to thank the following individuals for their review of this report: Keith A. Argow, President's Office, National Woodland Owners Association, Vienna, Virginia; and F. Stuart (Terry) Chapin, III, Institute of Arctic Biology, University of Alaska. Although the reviewers listed above provided many constructive comments and suggestions, they were not asked to endorse the content of the report nor did they see the final draft of the report before its release. The review of this report was overseen by Rodolfo Dirzo, Department of

Biology, Stanford University. Appointed by the NRC, he was responsible for making certain that an independent examination of this report was carried out in accordance with institutional procedures and that all review comments were carefully considered. Responsibility for the final content of this report rests entirely with the authors and the institution.

Contents

1

Introduction

The forested land in the United States is an asset that is owned and managed by federal, state, and local governments, families,[1] and other private groups, including timber investment management organizations and real estate investment trusts. More than 10 million family forestland owners manage the largest percentage (35 percent) of the nation's forestland acreage and the majority (62 percent) of its privately owned forestland (U.S. Department of Agriculture, 2012). The Forest Service of the U.S. Department of Agriculture (USDA), which is responsible for the stewardship of all of the nation's forests, has long worked with private owners of forestland on forest management and preservation, but all forestland is facing intensified threats because of the long-term effects of global climate change. The Forest Service recognizes that family forestland owners play a key role in protecting forestland and is working to identify optimal ways to engage this diverse group and support them in mitigating threats to the biologically diverse land they own.

REPORT FOCUS AND SCOPE

The task of engaging with family forestland owners is complicated by the fact that they receive technical, financial, and educational assistance

[1]Family forestland owners are defined by the U.S. Forest Service as families, individuals, trusts, estates, family partnerships, and other unincorporated groups of individuals who own forestland.

from a wide variety of public and private foresters, who themselves need to have a better understanding of how to engage forestland owners to deal with climate change. In addition, to effectively convey forest and climate change information in a manner that motivates action by family forestland owners necessitates not only a strong grasp of the science related to forest ecology and climate change but also an understanding of social, psychological, and educational theories on motivation, behavior change, and communication about complex issues.

The Forest Service and the National Institute for Food and Agriculture (NIFA) of the USDA asked the National Research Council's (NRC) Board on Science Education and Board on Environmental Change and Society to hold a public workshop, the fourth in the Climate Change Education Roundtable series, to explore approaches to the challenge of preparing the professionals (state foresters, extension agents, private forestry consultants, and other service providers) who advise or otherwise intersect with private family forestland owners on how to take climate change into consideration when making decisions about their forests. The Forest Service and NIFA were particularly concerned that most of these intermediaries lacked relevant expertise from the social and behavioral sciences and adult education. In response, an eight-person planning committee—the Steering Committee on Engaging Family Private Forest Owners on Issues Related to Climate Change—with expertise in human behavior, natural resource economics, risk assessment, communication, outreach and extension, and political and social sciences was assembled to plan and convene the requested public workshop. This two-day workshop, held in August 2013, focused on ways that findings from the behavioral, social, and educational sciences[2] can be applied in engaging private individual, family, and community forestland owners in conversations about preparing for the impacts of climate change. The steering committee was charged with addressing the following:

- Threats to forests posed by climate change and human actions.
- Private forestland owners' (individual, family, and community forest owners) objectives, values, knowledge, and dispositions about forest management, climate change, and related threats.
- Strategies for improving communication between forestland owners and service providers with respect to forest management in the face of climate change.

The workshop provided an opportunity for Cooperative Extension

[2]For convenience, these disciplines, which include economics, sociology, psychology, and political science, are generally referred to as the "social sciences" in this report.

foresters,[3] program administrators, nonprofit organization leaders, and others who are in a position to influence those who own and manage forests to interact with researchers and to reflect on how research can support strategies and tactics for effective communication and engagement with forestland owners. (See Appendixes B and C for the agenda and a list of participants.)

This summary, which was prepared by rapporteurs, describes the presentations and discussions that took place at the workshop. The views expressed in the report are those of individual workshop participants and do not necessarily represent the views of all workshop participants, the planning committee, or the NRC. The summary begins, in this chapter, with an overview of the ways in which U.S. forestland is likely to be affected by global climate change and other factors. Chapter 2 discusses the characteristics of the private owners of U.S. forestland. Chapter 3 discusses examples of strategies regional and local groups have used to engage with forestland owners. Chapter 4 discusses ideas drawn from the social sciences that can support productive engagement with forestland owners. Chapter 5 discusses ideas, outstanding questions, and possible next steps that emerged from discussion throughout the workshop.

CHALLENGES TO FORESTLAND

James Finley, professor of forest resources at The Pennsylvania State University, and David Cleaves, climate change advisor to the Chief of the U.S. Forest Service, set the stage with reflections on the challenges facing forests and their owners. Landowners value healthy forestland and want to protect it, Finley observed. Forestland owners' respect for the value of their land is a "logical platform" from which to address this group about the threats posed by climate change and human action, in his view. Healthy forestland is "an economic, an ecological, and a social value" that is critical not only for the private forestland owners who control and hold that land right now, he stated, but also for all U.S. residents, who will continue to benefit from the existence of that land.

Cleaves explained that the U.S. Forest Service and the USDA, and all other institutions that serve the forest sector, are eager for guidance from the social sciences in responding not only to a changing climate but also to demographic changes among forestland owners and other changes in the ways people interact with forests. Dynamic forests are part of the solution to climate change, he noted, because "if you want more carbon sequestration, you have to have healthier forests. . . . To get healthier forests, you

[3]The Cooperative Extension is a nationwide structure for providing information to agricultural producers and others; see http://www.csrees.usda.gov/Extension/ [November 2013].

have to understand the system—you can't just write a policy and turn it on and see what happens."

James Vose, research ecologist with the U.S. Forest Service, provided a detailed look at how forests in the United States are changing. The science of climate change is progressing rapidly, he pointed out, and new information is constantly being added to the climate assessment process. He based his presentation on the USDA report *Effects of Climatic Variability and Change on Forest Ecosystems: A Comprehensive Science Synthesis for the U.S. Forest Sector* (U.S. Department of Agriculture, 2012), which summarizes current research and discusses adaptation and mitigation strategies.

Drawing on that report, Vose looked first at how climate change will affect forest ecosystems in the United States. Forests develop in response to their physical environments, biological dynamics, and human decisions about their use, he explained. Ecosystems, climate stressors, and human actions are highly interconnected, so climate scientists use the word "anthropocene" to describe the world climate and all of its influences.

The changes that threaten forests begin with substantial increases in the emissions of greenhouse gases (see Figure 1-1), which have led to an increase in average overall global temperature of 1.4°F over the past 100 years, Vose explained. This increase has meant a significant loss of Arctic ice, an 8-inch rise in sea level since 1870, and an increase in heat waves and extreme temperatures during the past 20 years. The 2012 USDA report focused on using observations of how ecosystems have already responded to observed increases in temperature and other changes to provide some guidance as to what effects are likely in the future.

Predictions that were made 20 to 30 years ago, he explained, have largely been confirmed. At the same time, a number of extended experimental studies have now been operating for long enough that their results also reinforce climate scientists' capacity to make predictions about the future. The most widely used tools for making these predictions are forecasting models, such as the General Circulation Models or Global Climate Change Models, Vose noted. These types of models depict the range of possible scenarios and outcomes that can be expected under different sets of assumptions about such factors as carbon dioxide (CO_2) emissions, economic conditions, and resource use.

Modeling forecasts, together with the growing body of empirical evidence, indicate that climate change will have significant effects on U.S. forests, Vose explained, although different assumptions about CO_2 emissions, economic conditions, resource use, and other variables yield somewhat differing projections. Researchers in this area generally incorporate minimum and maximum estimates into their calculations to arrive at a realistic predicted range of outcomes. Throughout his presentation, Vose used estimates and predictions that fell in the mid-range of changes

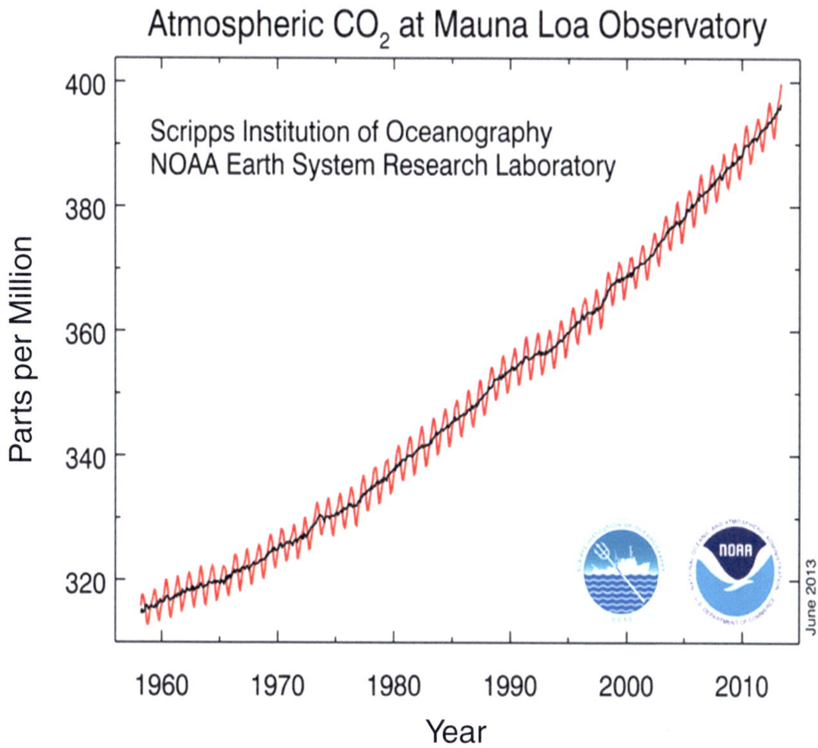

FIGURE 1-1 Change in atmospheric CO_2.
SOURCE: Available: http://www.esrl.noaa.gov/gmd/ccgg/trends/ [May 2014].

that could occur, between the extreme ends of the continuum of possible scenarios.

The direct effects will include increased temperatures, changes in the amount and extremity range of precipitation (which will mean more frequent and more severe droughts for many parts of the world), elevated CO_2, and a rise in sea levels. For forests, these changes will mean higher rates of growth where nutrients and soil moisture are available; reduced growth in forests with more limited water or that are affected by other disturbances; higher mortality of vegetation in drier areas; changes in habitat that will affect the distribution of plant and animal species; and changes in the hydrologic, nutrient, and carbon cycling processes.

Indirect effects that come with these changes with particular significance for forests will include increased risk of fire, changes in insect and pathogen populations, and growth of invasive species. Combinations of stressors that interact with one another, known as "stress complexes,"

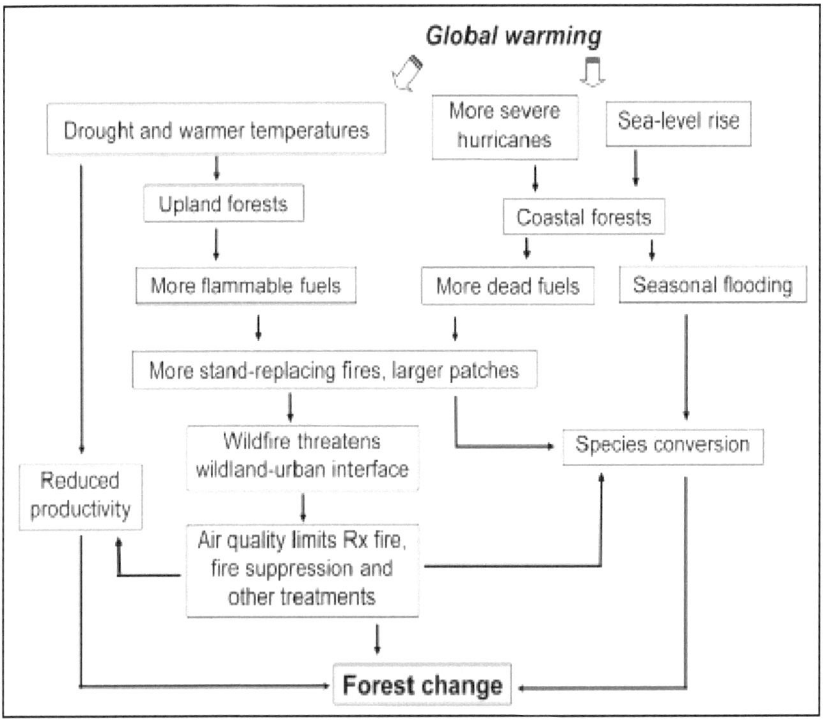

FIGURE 1-2 Stress complexes for upland and coastal forests in the southern United States.
SOURCE: Vose et al. (2012).

will be affected by a warmer climate; Figure 1-2 illustrates how this sort of interaction may accelerate changes, using the example of forests in the southern United States.

The distribution and abundance of tree species will change, Vose added, because changing conditions will substantially alter many habitats and make them unsuitable for the species that currently grow in them. There will be what he termed losers and winners: Some species may need to move north by as much as 400 to 800 km (250 to 500 miles), and the habitat changes may occur faster than species can migrate; other species' habitats may expand. Hydrologic and biochemical cycling will also change, in response to the combined effects of the changes in climate, Vose explained. The results for forests will include changes in snowfall and snowmelt, and increases in flooding, erosion, and landslide potential. Eastern forests are likely to store greater amounts of carbon—a benefit that may be offset by loss of forestland, drought, and other disturbances—while western forestland will store less carbon.

Vose emphasized that while the precise nature of the changes will be specific to particular regions, the big picture is one of "multiple, co-occurring stresses and disturbances that are likely to be much more severe in combination than if any of them occurred singly." Climate change is a large-scale phenomenon, he added, but the specifics in particular regions will influence the management response. Essentially, he said, the two management responses are to mitigate the changes and to adapt to them.

MANAGEMENT RESPONSES

The primary means of mitigating the harm, Vose explained, is to manage—slow down—the release of carbon into the atmosphere. U.S. forests capture about 16 percent of the carbon emissions from U.S. fossil fuel use at present, he noted, and "there is certainly opportunity to influence that through policy and action." Helping forestland adapt to inevitable changes by preparing systems and species, however, is the area where private forestland owners can play the largest role. Cleaves, in discussion, highlighted this point, noting that the USDA stresses retaining forestland, restoring function to forests with problems, and regenerating forests that have been depleted.

Figure 1-3 shows the distribution of privately owned forestland in the United States, and illustrates that in the eastern United States, the majority of forestland is privately owned. Thus, particularly in that region, private decisions about land use—whether to preserve it as forest or use it for another purpose, and whether to manage it—will play a critical role. (See Box 1-1 for the U.S. Forest Service's statement of its approach to forest management.)

Projected growth in human population will vary across the United States, Vose noted, with the most intense growth clustering around urban areas. The growth projected by 2060 will mean that significantly more land will move into the urban land use category, and this will happen at the expense of cropland and forestland. The number of U.S. acres that were forestland reached its lowest point in the 1970s, Vose noted, and has increased since then. Current projections indicate that by 2060, that number will again be at roughly the 1970 level, with most of the loss occurring in the private forestland of the eastern United States.

While much of this picture is quite daunting, Vose concluded, "Management activities can really influence the severity and the direction of the response." Landowners and forest managers, he added, are already accustomed to dealing with such external stresses as wildfire risk, insect outbreaks, and diseases. The pace and magnitude of climate change will

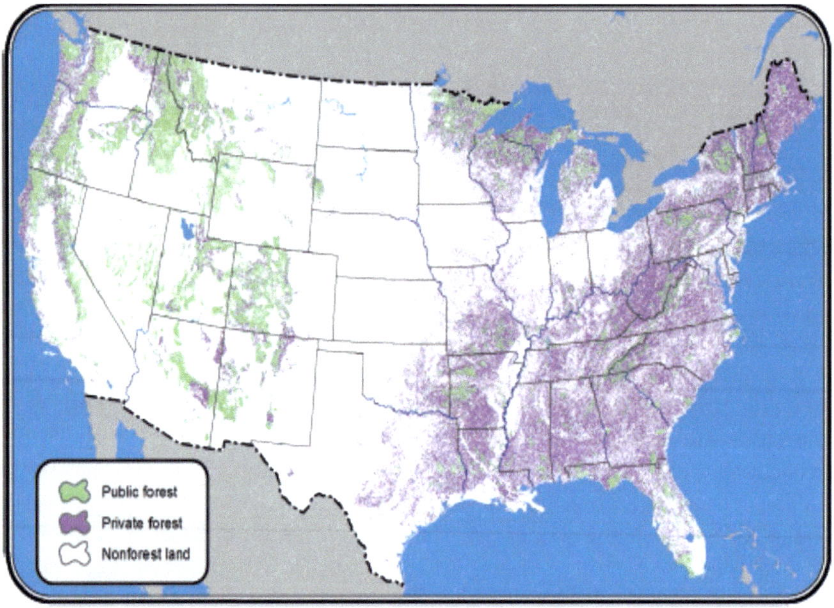

FIGURE 1-3 U.S. forestland ownership.
SOURCE: Walthall et al. (2012, p. 99).

BOX 1-1
The U.S. Forest Service's Approach to Forest Management

The overriding objective of the Forest Service's forest management program is to ensure that the National Forests are managed in an ecologically sustainable manner. The National Forests were originally envisioned as working forests with multiple objectives: to improve and protect the forest, to secure favorable watershed conditions, and to furnish a continuous supply of timber for the use of citizens of the United States. Forest management objectives have since expanded and evolved to include ecological restoration and protection, research and product development, fire hazard reduction, and the maintenance of healthy forests. Guided by law, regulation, and agency policy, Forest Service forest managers use timber sales, as well as other vegetation management techniques such as prescribed fire, to achieve these objectives. These activities have captured substantial public attention, and in some cases, become hotly debated issues.

SOURCE: Available: http://www.fs.fed.us/forestmanagement/aboutus/index.shtml [January 2014].

call for similar management responses, but with greater intensity and at a greater scale. The optimal management response will depend on the local circumstances, and will depend on "a lot of give and take, partnerships, and a lot of learning," he added. Scientists, landowners, policy makers, and land managers will need to work together to develop management plans, Vose said. But, he added, forestland owners will need to recognize that a tipping point[4] has been reached, and that "they could lose their forest as they know it in a very rapid sequence of events—wildfire, disease, drought, etc."

Vose closed with the observation that "the amount of private land in the United States means that to keep forest ecosystems healthy, productive, and providing the services that society depends on, the private landowner [will be] critical—this can't be addressed just on public lands."

[4]A tipping point is when there is a change from one stable state to another stable state; after the new state occurs, it may be impossible to return to the initial state.

2

Understanding Forestland Owners

From the picture of the threats to U.S. forestland and the challenge those threats pose for private forestland owners summarized in the previous chapter, the conversation turned to the characteristics of these landowners. Moderator Shorna Broussard Allred, associate professor at Cornell University, posed several questions: What drives their decisions? What sources of information do they draw upon to support the decisions they must make? Four presenters discussed what is known about forestland owners, their attitudes about and understanding of climate change, and strategies for engaging them in forest management.

PROFILE OF FORESTLAND OWNERS

Brett Butler, research forester with the U.S. Forest Service, presented recent results from the National Woodland Owner Survey, a census of forest owners conducted by the U.S. Forest Service.[1] This survey asks forestland owners questions about their land and how they use and manage it, and about their own characteristics and their goals with respect to their land. The survey was conducted in 1993 and 2006, and then in 2011, 2012, and 2013. The Forest Service expects to administer it again in 2015. Butler presented preliminary 2011 results because the team was still processing the more recent data at the time of the workshop. The unpublished

[1]The survey is part of the U.S. Forest Service's Forest Inventory and Analysis Program; for more information, see http://www.fia.fs.fed.us/nwos/ [January 2014].

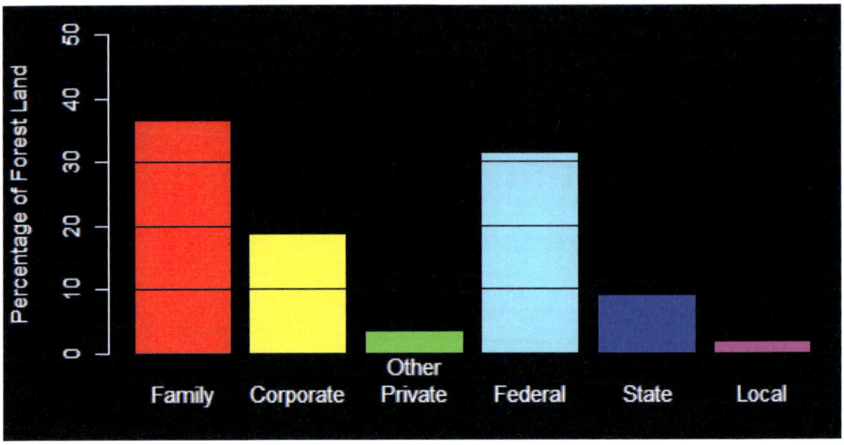

FIGURE 2-1 U.S. forestland ownership by class, 2011.
SOURCE: Available: http://www.fia.fs.fed.us/nwos [May 2014].

results discussed at the workshop do not diverge significantly from those reported in 2006 (Butler, 2006).

Butler's first point was that "forest owners rule," in the sense that it will be essential to see forestland through the eyes of those who own so much of it. More than half of the forests (by acreage) in the United States are privately owned, and two-thirds of those owners are individuals, families, communities, and other non-incorporated groups (typically referred to as *family forestland owners*), as shown in Figure 2-1. The majority of private forestland owners own between one and nine acres,[2] Butler noted, but many of the smallest parcels are actually urban and suburban land. Thus, a relatively large proportion of owners are associated with small parcels (particularly because many parcels are owned by couples and families), while the majority of forestland is rural (held in units of 10 acres or more). Among owners with more than 10 acres of forestland, the average holding is just under 60 acres. It is important to focus on both owners and acreage, Butler noted. The forestry community lacks tools and policies for working with suburban forestland owners, Butler observed, but is better equipped to deal with urban and rural owners.

According to the survey, owners with 10 or more acres have varied reasons for owning forestland. About 66 percent of owners, accounting for 57 percent of the land owned by this group, live on their forestland;

[2]See http://www.srs.fs.usda.gov/futures/reports/draft/Frame.htm [March 2014].

31 percent, accounting for 39 percent of the land, farm some portion of their forestland. These owners cite a range of reasons for owning their forestland, but their top five reasons are stable. While the ordering changed between 2006 and 2011, the top five reasons for those two years, respectively, were

1. beauty/scenery (unchanged);
2. part of home (2006), nature protection (2011);
3. privacy, family legacy;
4. nature protection, privacy; and
5. family legacy, part of home.

Butler noted that financial objectives did not come up in the top five, but the most frequently cited financial objective was investment (i.e., value of the land itself), not timber production (i.e., the productivity of the land).

The survey also provides information about the steps private owners take to manage their forestland. The most common management activities are cutting trees for personal use, removing invasive plants, activity related to wildlife, maintaining trails, and cutting trees for commercial use. Much of the private forestland is not being managed, not because the owners do not want to take care of their land but because they are not sure what to do, Butler explained. This situation "presents a big opportunity," Butler noted. Just 10 percent of owners, responsible for 23 percent of the forestland, report having implemented a management plan, and just 18 percent, responsible for 35 percent of the land, have received management advice. In Butler's view, given that most landowners do not have management plans, alternative approaches to engage with forestland owners are needed. He suggested that potlucks and other informal ways of spreading the word may be effective in helping private forestland owners know where to turn for advice.

Forestland owners have many concerns about their property, Butler explained. Although climate change is on the list, it is not among the top 10 concerns the owners cited, which are, in decreasing order of frequency,

1. Property taxes
2. Trespassing
3. Keeping land intact
4. Vandalism/dumping
5. Insects and diseases
6. Wildfire
7. Water pollution
8. Invasive plants

9. Wind/ice storms
10. Air pollution
11. Development
12. Off-road vehicles
13. Drought
14. Global climate change
15. Damage from animals

This window into the priorities of landowners reinforces the point that "we [scientists] need to couch our concerns in the context of their [private forestland owners] concerns," and to understand who they are, Butler suggested. As a group, private forestland owners are relatively old—about 40 percent are 65 or older, and of that group about half are 70 or older. Seventy-nine percent are male—though Butler noted that in many cases the owner is a couple—and 95 percent are white. Forty-eight percent have a college degree, and 27 percent have an annual income of $100,000 or more. Owners responsible for about 20 percent of privately owned land (in parcels of 10 or more acres) reported being moderately or highly likely to sell their land in the next five years, Butler noted. This raises important questions, he noted: "Who will the next generation be? Will they have the same desires and concerns? Will we communicate with them differently?"

TARGETING FORESTLAND OWNERS

Mary Tyrrell, executive director of the Global Institute of Sustainable Forestry within the Yale School of Forestry & Environmental Studies, built on this profile, describing the results of research that uses social marketing techniques, statistical analyses, and focus groups to explore forestland owners' perspectives and views on climate change.[3] First, she described four basic categories of owners, who differ in how they use and view their land, in their value systems, and in the sorts of messages that are likely to reach them.

One group (40 percent of owners) uses the land as a woodland retreat and is motivated primarily by stewardship. This group wants to protect and enjoy the forest's beauty and wildlife. They tend to own a below-average acreage and account for 35 percent of privately owned forestland. "If you are trying to reach these folks," Tyrrell explained, "you should talk about healthy woods, wildlife, conservation, recreation, and how to protect their land for the future."

[3]For more information about the research, see http://www.engaginglandowners.org and http://www.sustainingfamilyforests.org [May 2014].

A second group (30 percent, accounting for 37 percent of the land) works the land—primarily as farmers who have wooded lots. This group also uses the land for recreation, including hunting. Individuals who belong to this group share some of the broader interests of the first group, but they also appreciate the financial value of their land, which is as important to them as stewards or the forestlands' ecological health, Tyrrell noted. They share an ethic of respectful and judicious land use and will also respond to messages about protecting financial health (such as timber harvests that leave the land in good shape), and recreational uses.

A significantly smaller group (8 percent of owners, accounting for 12 percent of the land) uses their forestland for supplemental income (e.g., through timber harvest) or as an investment in land. This group is particularly interested in keeping the land intact for their heirs and in possibly increasing its value. Consequently, members of this group tend to be interested in reducing taxes and other liabilities, in cost-share and other incentive programs that increase financial returns, and in professional advice about managing and protecting their land.

The last group is described as "uninvolved" because they view the land as a general investment and prefer not to expend much time or attention on it. Twenty-two percent of the owners, accounting for 16 percent of the land, fall into this group, Tyrrell noted, but they do not tend to participate in focus groups. These are owners who do not report having strong values associated with their land and are likely to live elsewhere, or to be farming neighboring land but have minimal interest in the forested portion. Although there is little information about this group, Tyrrell speculated that they might be interested in messages about increasing their land's value and keeping it intact for their heirs.

GOALS FOR EXTENSION PROGRAMMING

Janean Creighton, associate professor and extension specialist in the College of Forestry at Oregon State University, described research on what forestland owners in Alaska, Idaho, Oregon, and Washington know about climate change, the impacts of forest management on forest resilience to climate change, and how extension programs can best meet their needs. The work began in 2008 as a needs assessment, she explained, when forestry extension leaders in the region wanted to develop programs on climate change for private forestland owners. The research included 24 focus groups with 193 forestland owners who had participated in extension programs in the four states, and whose territory includes four distinct types of forest, as shown in Figure 2-2.

Some "very, very lively discussions" developed during the focus group sessions, Creighton explained, and it became clear that the concept

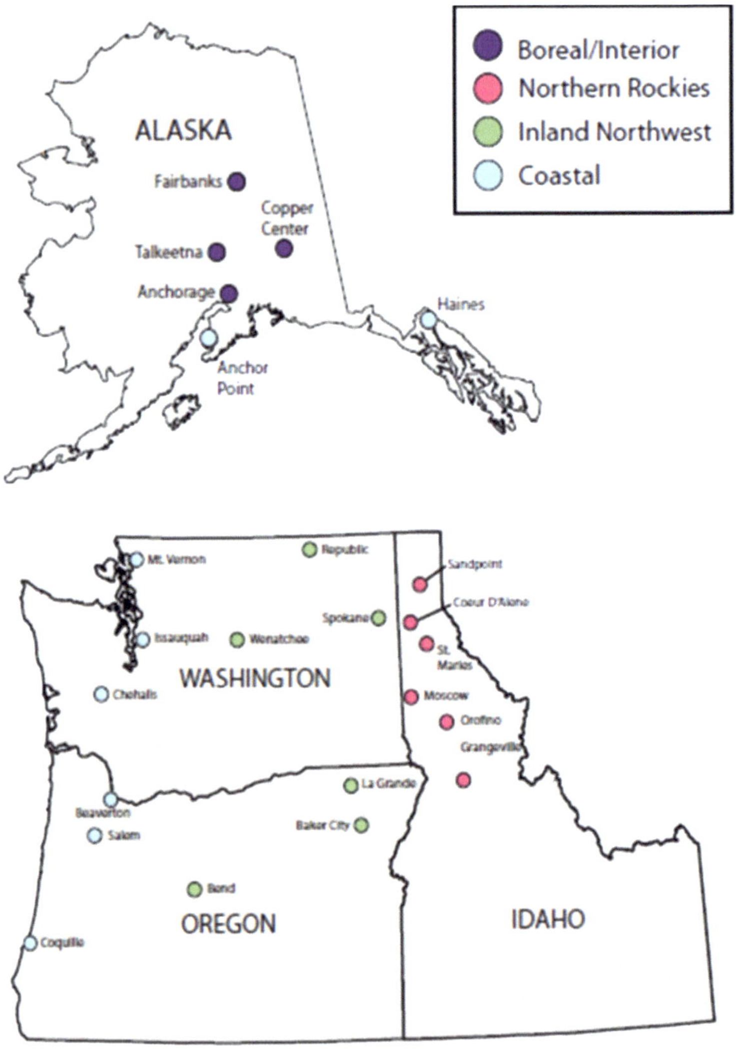

FIGURE 2-2 Northwest and Alaskan forestland categories.
SOURCE: Grotta et al. (2013).

of climate change was very value-laden, political, and emotional for the individuals who participated. The participants reported that they received their information about climate change primarily from the media. This included news outlets from Fox News to NPR, Creighton noted, but the information was "usually very passively gathered." Although some of the participants were scientists or reported getting information from extension programs or other university experiences, the group as a whole reported significant skepticism about science, she added. She noted that the information was collected in 2009—and owners' thinking may have changed since then—but that at that time, many of the participants perceived that the scientific community could not agree, and that much research may be influenced by financial incentives and politics.

In general, this group—with the exception of those living in Alaska—did not believe that climate change was affecting their forests. The owners were accustomed to dealing with problems such as insects, fire, disease, and the like, she noted, but they did not perceive those problems as being connected to climate change. Owners in Alaska, however, reported that they were seeing direct effects of climate change, including loss of permafrost and receding glaciers, and that these changes were happening rapidly—much more so than any changes perceptible in the Pacific Northwest.

The forestland owners reported during the focus groups that they had made few or no changes in their forest management practices because of climate change. As in the past, Creighton explained, they were interested in managing to preserve the resilience of their forestland, maintain biodiversity, and protect its beauty and value. They said they were very interested in practical information and management advice that was relevant to their region.

A few common themes emerged from this work, Creighton suggested. First, she and her colleagues concluded that the forestland owners were not uninterested in learning more about climate change, but rather skeptical, though willing to be convinced. For the owners, the issues of most concern in terms of forest management reflected a shorter time horizon than that for the projected effects of climate change. They worry about timber harvests, the need to replant to replace lost trees, and so forth, rather than more distant threats, she explained. They indicated that they did not have enough information on local impacts to suggest changes they should make. They perceived the effects of climate change as very gradual, Creighton explained, and they "thought there would be time enough to respond as they saw things changing over time." The owners were very mindful of risk, she added, and were reluctant to "change practices now to adapt to some scenario that might occur in the future."

Creighton and her colleagues developed several suggestions for extension programming based on these findings:

- The forestland owners told the researchers that they would like help sorting through the competing claims they saw in the news, and Creighton and her colleagues suggested that *transparency and education with respect to the way scientists conduct and fund their work* and submit it for review by their peers would help the forestland owners better understand the issues.

- *Information framed in the context of local conditions*, including both the specific nature of the forestland and local cultural traditions associated with the forest, would be most meaningful to the landowners.
- Programming could help forestland owners *increase their understanding of modeling* and the range of uncertainty in scientists' projections, because these concepts are critical to understanding climate change.
- Programming could build on what the forest management owners are already doing to *increase adaptation.*
- Programming could *directly address forest policy and its impacts*, to counter perceptions that climate change policies themselves can be a threat to landowners' interests.

Geoffrey Feinberg, a research specialist at Yale University's Project on Climate Change Communication (YPCCC), also discussed research on views of climate change, based on work conducted by the YPCCC. The YPCCC explores what the public knows about climate change, how great a risk they believe it to be, and what drives their decision making related to it. The YPCCC also designs and tests strategies for engaging the public in climate science topics and solutions.

Feinberg began with an overview of Americans' views on climate change. The majority of Americans believe that global warming is occurring, though the percentage that do so fluctuates, as Figure 2-3 shows. However, only half of survey respondents believe that climate change is being caused by human activity and that percentage also fluctuates (from 57 percent in late 2008 to 49 percent in early 2013); one in three survey respondents believes that while climate change is happening, it is due to natural causes. Only 42 percent of Americans know, Feinberg added, that scientists are virtually unanimous (97 percent) in thinking that humans are causing climate change; 33 percent believe that there is significant disagreement among scientists. This is important, Feinberg noted, because "our research indicates that if you can get people to understand that climate scientists agree that climate change is happening, that is the gateway to getting them to understand some of the other important things about climate change."

In surveying the public about their views about climate change,

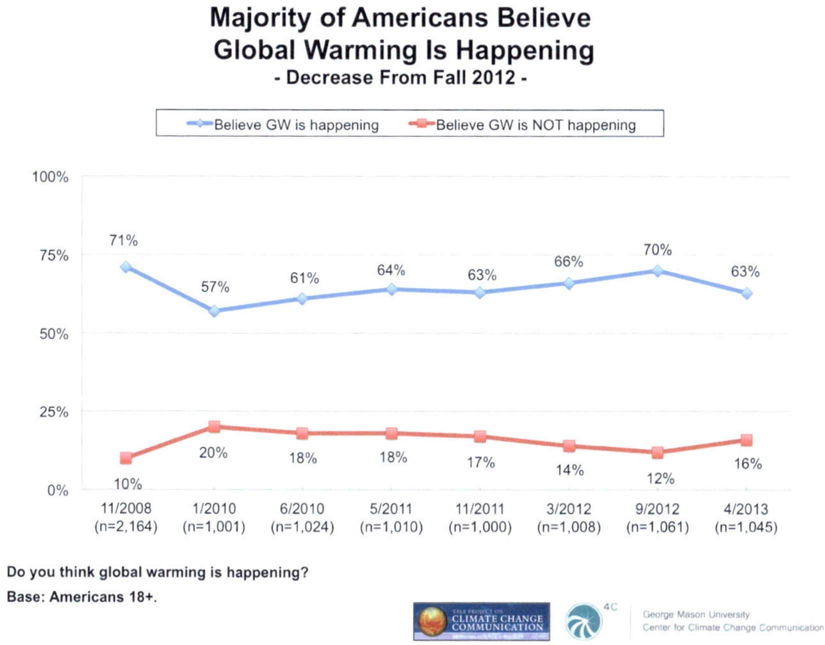

FIGURE 2-3 Perceived reality of global warming.
SOURCE: Leiserowitz et al. (2014).

Feinberg and his colleagues have identified six categories of people, which they call the "Six Americas."[4] These categories are as follows:

- **Alarmed (16 percent).** This group knows that climate change is caused by humans and believes that it is an urgent and serious threat. They have taken personal action to help mitigate climate change, such as recycling and buying hybrid cars. Demographically, they tend to be women, middle-aged, college educated, and moderate to liberal politically.
- **Concerned (26 percent).** This group also knows that global warming is real and serious, and they agree that it requires policy intervention. They are somewhat less involved than the *alarmed* group and less likely to take steps to reduce their own carbon footprint. Demographically, they resemble a cross-section of Americans.

[4]See http://environment.yale.edu/climate-communication/files/Six-Americas-March-2012.pdf [February 2014].

They tend to have egalitarian values and are fairly moderate politically.

- **Cautious (25 percent).** This group views global warming as a problem, but not as an urgent or near-term personal threat—they feel less urgency to act than the first two groups. The *cautious* group includes both Democrats and Republicans, but they are not very civically engaged.
- **Disengaged (5 percent).** This group is not interested in the issue and does not know much about it, but they are the most likely of all to say they could change their minds about it. They are comparatively less educated than the other groups, and have a lower average income. Minority women are over-represented in this group.
- **Doubtful (15 percent).** This group is undecided—some believe that global warming is happening and some do not, but most think that it is not caused by human actions. They do not view it as an imminent threat. They are older than the other groups and more likely to be white. They are predominantly fairly well-educated Republican men, and they have strongly individualistic values.
- **Dismissive (13 percent).** This group believes very strongly that climate change is a hoax and is actively engaged and vocal on the issue. This group is predominantly male and politically conservative. They are politically active and strongly oppose a policy response to climate change. They hold firm traditional religious beliefs and reject most government intervention.

Feinberg closed with the suggestion that learning how the four types of forestland owners described by Tyrrell are situated within this typology of "Six Americas" would make it easier to tailor messages to the owners. For example, he suggested, if one knew that many of the owners who use their forestland for supplemental income were in the *cautious* or *doubtful* categories, it would make sense to focus on the message that renewable energy is valuable because it will help make the United States energy independent, rather than messages about saving the planet and endangered species. While communication efforts with the *dismissives* probably will not be particularly fruitful, he opined, approaching owners with messages that reflect their concerns is most likely to be successful.

DISCUSSION

Questions and discussion touched on several of the issues raised in the presentations. One participant was concerned about the lack of understanding of the science of climate change and how that can be addressed. Mary Tyrrell suggested that the best approach may be to simply leave climate change out of conversations with forestland owners. The owners are already concerned about many of the problems that are likely to increase as the climate changes, such as invasive species. It may be just as effective to talk to the owners about what to do about these problems specifically as to push education about climate change, in her view. She added that "if you can get landowners working together on landscape-scale issues, maybe you can achieve something on the ground without having to have a battle about whether they are going to believe that climate change is happening or not."

Janean Creighton agreed, but noted that there is disagreement among some extension professionals about this, because leaving climate change out of the discussions is not completely honest. Some people are convinced that "you need to let people know what the long-term problem is and that is climate change," she observed. On the other hand, Geoffrey Feinberg noted, "there is some research suggesting that education often simply serves to arm people with information they need to better argue their preconceived notions—so it may be best to approach people on the level of their concerns."

Another participant was concerned that talking with landowners and others about their beliefs with respect to climate change is a mistake. "Belief comes out of cultures and societies," the participant commented, "but in science, we accept or reject hypotheses," adding that calling acceptance of the science a belief may undermine efforts to address misunderstanding of the science. The panelists acknowledged the point but several observed that for some people, this issue is in the realm of belief. They may lack understanding of models and projections, but they associate climate change with other controversial natural resource topics that raise beliefs about, for example, the rights of animals or the relative value of endangered species. It may be necessary to address people in the language they are accustomed to in order to reach them, several participants and panelists suggested.

A related point concerned the issue of credibility. One participant noted that "there are some people who have adopted what we [extension agents] have been advocating for, but because of current marketplace issues, they cannot move forward." They are likely to distrust further recommendations if previous recommendations from extension agents had negative environmental or economic consequences. Similarly, many in the Pacific Northwest group who participated in the focus groups, Creighton noted, believe that the impact of regulations would make it even more difficult for them to manage their lands. They feel they are already experiencing severe restrictions and heavy regulations, for example, to protect salmon habitat and spotted owls, she explained. "They are concerned that there is going to be another public good that is recognized" and that will mean more control over how they manage that public good, she added.

3

Communicating About Forests and Climate Change: Case Studies

States, counties, cities, other jurisdictions, and private organizations around the country are engaged in a wide array of projects designed to engage forestland owners in protecting and managing their land. The focus on the impacts of climate change on the forestland varies greatly across these projects. Speakers described their experiences, discussed their approach to addressing climate change, and identified strategies they had found effective and ineffective. Their remarks focused on regional and community forest management, online extension services, engagement of forestland owners and local governments in forest management, and minimized threats to forest health and sustainability. Follow-up discussion with participants highlighted additional questions.

REGIONAL AND COMMUNITY FOREST MANAGEMENT

Seventy percent of New England is forested, and 86 percent of that forestland is privately owned. A considerable portion of those privately owned forestlands is being managed by hundreds of local and regional land trusts that share concerns about the forestlands' well-being, Bill Labich, regional conservationist at the nonprofit organization Highstead, explained. Thus, regional partnerships—informal networks of public and private agencies and organizations—are key to conservation in New

England, he observed.[1] He described how Highstead acts as the steward of a tract of communally owned land and also both conducts and sponsors ecological research. Highstead is dedicated to increasing conservation and sustaining the forested and natural landscape in the region, in part by helping groups coordinate their efforts and collaborate across political boundaries; it also collaborates with Harvard University in a New England-wide conservation effort called Wildlands and Woodlands.[2]

Five years ago, Labich noted, many New England conservation groups that focus on forestlands did not even know of other such groups in the region. Now, however, Highstead sponsors an annual meeting to bring the groups together and also provides technical training, funding, and supportive research. Support and activities include web resources (e.g., an actively facilitated LinkedIn group), webinars, and gatherings at the state and local level.

In the New York/New England Family Forest Owner Outreach Initiative, one such collaboration facilitated by Highstead, state foresters, university researchers, and conservationists from four states worked together on engaging family forestland owners in stewardship and conservation activities on their own land and in the surrounding landscape. The project began with collaborations to increase understanding of the people who own New England forestland, and to identify effective strategies for engaging forestland owners. Regional partnerships from each of the four states identified threats to their forests and potential solutions, Labich noted, as well as specific messages they believed would be most effective with forestland owners in their regions.

The partnerships reached out to owners they were able to find through statewide databases or their own research on local data. They used postcards that flagged questions the owners might have about, for example, harvesting and selling trees, keeping woods healthy, reducing property taxes, learning about wildlife habitat, or improving recreational activities on their land. The owners were invited to educational forums and introduced to additional programs serving their area, such as a bird habitat assessment that provided an opportunity to meet with a wildlife biologist to learn more about how their land can be managed as a bird habitat.

Labich identified a few lessons from this work. One key to its success, he explained, is that "there is a sequence of events." Owners become engaged and are guided to supports that meet their particular needs, while the partnership can track how the owners respond and evaluate

---

[1]Labich referred participants to the September 2013 issue of the *Journal of Forestry* for more information on regional partnerships.

[2]For more information, see http://www.highstead.net/about/history.shtml and http://www.wildlandsandwoodlands.org/ [January 2014].

the results of different efforts and programs. Evaluation tools include a landowner database, sign-in sheets for events, surveys of owners' attitudes, and follow-up phone calls. Learning what works is a primary goal of evaluation efforts, Labich explained, so Highstead collects feedback on its outreach efforts, but also data on subsequent actions that result from such efforts, such as meeting with a forester and developing a management plan or initiating a request for a land easement.

Labich emphasized that personal attention is very important: The postcards are far more effective, for example, if they are sent to owners by name, with a personal note, rather than by bulk mail. At the same time, owners who have not previously engaged with this sort of activity have welcomed opportunities to make personal connections and expand their networks within their own areas and states and also across state lines. The partnerships focus on the many benefits of connecting large forest blocks and increasing the resiliency of forestland, he added. They generally do not use climate change as the hook to engage people.

Bringing together communities that live near forestland is also a strategy used to create partnerships that benefit the health and sustainability of forests. For example, community forests—those that are not privately owned—are common throughout the world, noted Martha Lyman, a founding partner of the Community Forest Collaborative. Community forests range in size and structure, she added: They may be owned by a local government, a tribe, or a nonprofit organization such as a land trust or a conservation group, or be covered by a long-term stewardship contract with the Bureau of Land Management. They are supported by numerous organizations and institutions, including the Open Space Institute, the U.S. Forest Service's Community Forest Program, and, in the New England region, the Community Forest Collaborative.

Lyman explained that the Community Forest Collaborative, which operates in Maine, New Hampshire, and Vermont, was designed to take advantage of the resources and expertise of partners concerned with healthy forestland to promote community ownership and participation in forest management, to facilitate access to the benefits and value that the forest can bring, and to protect forestland generally.[3] The collaborative focuses on the well-being of the community, treating forestland as an asset to civic life and a source of economic opportunity, as well as a resource to be conserved.

Though conservation and protection are critical, Lyman noted, the Collaborative found that talking about investments, assets to the community, and the community's capacity for leadership was more effective in engaging communities and individuals than conservation itself. Two

---

[3]See http://www.northernforest.org/Community_Forest_Collaborative.html [January 2014].

projects illustrate how the Collaborative's work with communities has addressed issues associated with climate change, she added.

The small town of Randolph, New Hampshire, encompassed a 10,000-acre tract of wooded land. She explained that the forest changed hands several times in a few years and the town grew concerned about protecting it. Randolph's citizens were concerned not only about preserving the beauty of the tract located between two segments of the White Mountain National Forest, but also about the possibility that they would have to fund services for new development if the tract were broken up into parcels. With the assistance of the U.S. Forest Service and other partners, the town was able to buy the forest. The Forest Service was glad to protect an important wildlife corridor, she noted, and has fostered a very productive relationship in which resource and management strategies are shared. The town is very committed to long-term stewardship and has invested in monitoring stations that have been placed throughout the forest, and has invited researchers and visitors to help with gathering data. Lyman noted that public participation has been critical to success from the beginning of this project. Public hearings provided a venue for input and communication, and the public remains engaged in setting management priorities for the forest.

The town of Grand Lake Stream, Maine, faced the same concern when the Farm Cove Forest nearby changed hands repeatedly in a short time. The town, however, did not have the capacity to acquire the land, and also faced disagreements about forest issues. A group of citizens set up a private, nonprofit land trust that was able to acquire 27,000 acres. They later added an additional 6,000 acres, and the land became the Farm Cove Community Forest. Protecting this forest was important at a broader scale, Lyman explained. It, too, is in the middle of a wildlife corridor, as well as an international conservation initiative, the Downeast Lakes Forest Partnership.

These and other community forests slow the pace at which forestland is being fragmented, Lyman observed. Both serve as buffers for ecological reserves—a key component of adaptation to climate change. Many community forests are watershed lands, she added, so managing them in ways that protect water quality is important. They can also provide value to their communities in surprising ways. For example, the Farm Cove Community Forest raised $200,000 by selling carbon offsets, which allowed them to purchase an additional 21,000 acres. In these and many other cases, she added, collaborative management across ownership and public engagement are critical to success.

ONLINE EXTENSION SERVICES

Online communication has provided another vehicle for engaging landowners in climate change issues, as Eli Sagor, extension educator at the University of Minnesota, explained. He illustrated the possibilities for this approach with a website developed by the University of Minnesota, called *My Minnesota Woods*.[4] This site has multiple features, he noted, including a Twitter account, Facebook page, and email newsletter. The programming is focused less on the broad challenges of climate change—he noted that they rarely use the term—but rather on improving forest health. Thus, on *My Minnesota Woods*, adaptation rather than mitigation is addressed, and such goals as keeping forestland forested, increasing productivity, and keeping trees healthy are stressed. Extension educators use the website to encourage owners and others to monitor forestland; notice changes, opportunities, and threats to forestland; and inform themselves about how to interpret the changes.

Most owners do manage their land, Sagor observed, and the website's strategy is to encourage them to increase and improve the ways they manage their land. The website invites and facilitates dialogue, providing platforms for answering the questions owners are likely to have in an inexpensive and accessible way. It provides guides to reliable, research-based information about such questions as, "What are my basic forest management options? How does it work to sell timber in Minnesota? Who does what? How can I make sure it goes well? What forest health issues should I look out for?"

The information on the website is easily accessible by using everyday language and contains links to other web locations that offer workshops and other programming. Sagor and colleagues recognize that attending a workshop is expensive in both time and funds, and that many owners want to focus on the information they personally need. Therefore, they disseminate the workshop information in a variety of ways so that owners have access to what is relevant to them. They have also begun to expand their reach, for instance by collaborating with the Sustainable Forest Education Cooperative, which targets professionals who work with forestlands, such as loggers.

In Sagor's view, the focus on forest health has been effective in engaging forestland owners. Global perspectives on forestland management and conservation have been too remote or abstract to reach his target audience, and he believes the message about climate change is both frightening and confusing for them. On the other hand, owners can experience forest health; they can "see and touch and feel—they see the bugs in their

---

[4]See http://www.myminnesotawoods.umn.edu/ [January 2014].

woods and the trees dying." Forestland owners, he added, may not think of themselves as foresters, but "they love watching things grow; they love noticing the seasonal changes."

To leverage this interest in direct observation, *My Minnesota Woods* developed a program on phenological observation, in which participants report seasonal changes to flowering they observe to the National Phenology Database, using their smart phones or home computers. In this way, users build data about their own land, contribute to a national effort by collecting data in a manner consistent with standardized protocols, and also receive support in interpreting the data. The phenology observation program is an example of the ways extension educators associated with *My Minnesota Woods* try to package climate-related content together with content that already has a broader appeal. However, Sagor noted in response to a question, those who are engaged in local observations already report climate-related changes and are willing to engage in discussion of climate change science as a result of their first-hand experiences.

Sagor also commented that the interactive nature of the website is important in engaging the target audience. The website provides venues not only for users to ask questions, but also for them to post their own reflections on what they have learned and observed. Every page has a moderated comments feature in order to control for abusive posts. They also use a variety of methods to reach, and deliver content to, landowners who may not be thinking about forest management or the resources available through extension, including monthly e-newsletters with links to their own and other websites.

Sagor closed with a few thoughts about outcomes. The goal, he explained, is not only to increase traffic to their site and engage more users in their newsletters, workshops, and other resources, but also to increase user engagement with volunteer and leadership activities, for example, and influence their actions. To illustrate how this can work, he explained that 65 percent of the traffic on the Oak Wilt webpage comes from elsewhere on the web through an online widget that is posted on partnering sites and that allows web users to receive specific information from *My Minnesota Woods*. Forty-five percent of those visitors then go on to view other extension content. This is a small step, he acknowledged, that is "far removed from action on the ground," but they have developed a framework for tracking users' pathways, and also for identifying changes they make after participating in a program. "We try in a variety of ways to put the content where people look," he concluded. "We are not trying to convince the skeptics, but when someone wants to go looking they are going to find it—that is the low-hanging fruit: focusing increasingly on impact, measuring those conversions, understanding how these tools are helping us get where we need to go."

A project of the University of Arkansas's extension program illustrates another way to use internet-based resources to target messages to particular groups, explained Tamara Walkingstick, associate director of the Arkansas Forest Resources Center at the University of Arkansas. Arkansas is what Walkingstick described as a "traditional extension state." They have two county agents in each of the state's 77 counties, which mean they have "a massive network—it is easy to get information out there." By law, she added, every forester, or anyone who gives advice on forest management to forestland owners, is required to engage in eight hours of continuing education on the subject every year. A network of partners, including the Arkansas Forestry Commission and the Society of American Foresters, works together to provide this education. Most of the foresters prefer to acquire their annual credits in one day, so those workshops are a primary education opportunity. "We don't use the words 'climate change,'" Walkingstick noted, but it is very easy to incorporate some of those concepts into programming for "this captive audience."

The extension program is using additional methods to broaden their reach. In one example, the Invasive Species Education Project, they train forestland owners to become "first responders" who can identify new invading species that turn up in Arkansas forests. The training includes the day-long workshops and a website, and extension is developing an online course. They plan to add a statewide conference and other educational materials. Walkingstick and her colleagues encountered some resistance with respect to the online course, she noted, because some extension foresters were concerned that older forestland owners would not be comfortable with using technology in new ways. She acknowledged that may be true to some extent, but that they need to be prepared to reach younger people as well.

The Invasive Species program, Walkingstick explained, has reached approximately 80 percent of foresters who are registered to practice in Arkansas (400 individuals) who, collectively, manage approximately 3.3 million acres, or about 18 percent of the forest acres in the state. The program has also reached more than 250 Master Gardeners. The extension program uses an online survey to collect information about changes people make after using its resources, she noted. In her view, there are several keys to the success they have tracked so far. One is the partnerships they have established, and another is the energy they put into talking with the stakeholders they are trying to reach, to learn not only about their needs and problems, but also to benefit from their knowledge of local conditions.

ENGAGING FORESTLAND OWNERS AND LOCAL GOVERNMENT IN FOREST MANAGEMENT

Allyson Muth, a program associate in the Forest Stewardship Program at The Pennsylvania State University (PSU), described how the program connects forestland owners by training them to engage in extension efforts.[5] The program includes a peer volunteer network, which, Muth explained, has been a useful way to introduce owners to difficult topics. The primary goal of the Peer Volunteer Program is to train woodland owners to better manage forestland and to motivate others to be stewards of forestland. The program is similar to the Master Gardener Program that many extension services have: Participants receive training over two weekends in basic forestry principles and in exchange they agree to care for their own forestland and to share what they have learned with others.

Pennsylvania has approximately 700,000 forestland owners, who own their land, on average, for 18 years. "That is a lot of turnover," Muth observed, "and a lot of people to reach." By engaging these owners in sharing what they have learned, she added, the program has a "multiplier effect." This approach, in which forestland owners speak from their own experience to other forestland owners, also draws on common models of adult learning. When adults encounter new information, she explained, they tend to evaluate it in terms of how it fits with what they already know and have experienced. They may reject new information if it is in conflict with their prior understanding. Trust is also a key to acceptance of new information, she added. Research (Gootee et al., 2010) has shown that forestland owners assess the caliber and trustworthiness of new information in part based on the degree of comfort they feel with the person delivering the information, she noted. Forestland owners, like most adults, are most accepting of information that comes from someone who is like them and has had similar experiences. They may believe that information coming from a person with a degree in forestry or wildlife biology will not be accessible or useful to them, she suggested.

The program has trained more than 550 volunteers since 1991, Muth noted, and 430 are currently active. Some volunteers contribute considerable hours—in 2011, the group collectively volunteered for the equivalent number of hours that 14 full-time employees would work, reaching 36,000 landowners. This is approximately the same number that paid extension employees were able to reach, she added.

The PSU Forest Stewardship Program has focused on making sure that messages about climate change are effective, given that previous mes-

[5]See http://extension.psu.edu/natural-resources/forests/private/stewardship/pennsylvania-forest-stewards [January 2014].

sages that directly address climate change have not been well accepted by forestland owners. They have conducted statewide surveys of landowners, reaching 100 owners in each of 66 counties. They found that owners in Pennsylvania cite reasons for owning their land similar to those found in the work described above by Butler (see Chapter 2):

- Enjoying and sustaining wildlife–62 percent
- Experiencing solitude–59 percent
- Incidental ownership (the forest came with a home or farm)–49 percent
- Engaging in recreational activities–34 percent
- Using wood (mostly for burning)–38 percent
- Investing into land (in part as family legacy)–14 percent
- Harvesting timber–13 percent

Follow-up interviews and focus groups have reinforced the idea that financial objectives are not the primary concern for most forestland owners. Many have "a deep and emotional connection to the land," Muth explained. "It reflects their relationships with other people—and sometimes protects them from others. It reflects their relationship to time, to the world, to their bodies." Extension has taken this information seriously, Muth explained, adding that "we have gotten brave and followed them to the touchy-feely kind of thing, because owning land is not a logical proposition—it is a headache. It is a heart decision, not a head decision."

She described how extension works to help owners articulate their own reasons for valuing the land—sometimes for the first time. This process helps extension and, in turn, guides the owners in evaluating their options, making decisions, and planning for the future. Extension focuses on good forest stewardship and the forest legacy that owners will leave for future generations. This topic has been a platform for talking about keeping land forested and mitigating change. The long-term focus has also meant supporting owners in thinking about estate planning, because turnover is often a threat to forested land. Many landowners have thought about plans for their land but fewer have written formal plans. Thus, Muth explained, the extension program conveys to the owners that good forest management means planning for changes that may come after the current owner is gone. Most landowners want to do the right thing, she added, but they do not always know what that is (Jones et al., 1995). Relating the idea of climate change to the responsibility to be a good steward appears to be helping to make difficult messages more palatable, she added.

Counties and local governments can also have a significant influence on the forestlands in their jurisdictions, explained Don Outen, natural resource manager with the Baltimore County Department of Environmental Protection and Sustainability. There are nearly 3,100 counties and county equivalents in the United States, with heterogeneous vegetation and ecosystems. Some have a lot of forestland and few people; in others, the reverse is true. However, because of their planning and zoning authority, counties and local governments can have significant influence on forest resources. It is vital to engage them in the goal of keeping forests forested, in Outen's view.

Baltimore County has had success for more than 45 years in managing growth, he observed. Although the county has very densely populated areas, "nearly half is tree canopy." The department is not staffed with researchers, he noted, but they have focused on conveying to landowners research-based ideas about water quality and keeping forestland healthy. More recently they have folded climate change into the basic message that "all values are at risk" when forest health is threatened. Sustaining and increasing healthy forests and trees, he noted, is the most cost-effective tool for mitigating the effects of climate change. Baltimore County has identified four priorities for its forests: keep forestland as forest; strategically re-forest stream buffers, areas adjacent to existing forests, and urban areas; restore and maintain forest health; and provide incentives for and training in stewardship to the citizens who own 75 percent of the county's forest cover.

In practical terms, this has meant such efforts as supporting local jurisdictions in complying with federal mandates related to water quality. "Clean water is rooted in forests," he noted. "The forests are part of a strategic tool for complying with these mandates." Outen explained that the county has at its disposal a range of tools and partnerships to address healthy forests. It has a Forest Sustainability Program, which grew out of its participation in pursuing the goals of the Montreal Process,[6] an intergovernmental effort to promote sustainable forest management. The county is also part of the National Roundtable on Sustainable Forests and has been working with numerous organizations to build interest in the message about forest sustainability.

Baltimore County emphasizes the public interest in healthy forestland and has collected data on the vulnerabilities of local forestlands. Identifying the relative value of different parts of the forest, for example by identifying the patches that cover the greatest percentage of the stream system, has been important, as have data on the very fragmented nature of forestland ownership. Outen opined that the number of owners in the county is significantly higher than in most regions because of small

---

[6]For more information, see http://www.montrealprocess.org [January 2014].

average parcel size: Approximately 35,000 people own more than 9,000 forest patches in the county.

A considerable issue in addressing forestland management issues is that many owners do not even recognize themselves as forestland owners, Outen noted. The county has used "large-lot, low-density development as a part of its growth management strategy," he explained. There are many lots of five or fewer acres that are wooded. The owners may report that "I didn't want all this land—it just came with the house," he noted. The land may have tremendous potential, but the owners do not know what species are growing there or recognize invasive ones. Outen and his colleagues have found that these owners "don't know where to begin and are paying a horrendous amount of money to mow their open land because they don't know what else to do with it."

In response, the department piloted a program called Turf to Trees, in which they used Geographic Information Systems (GIS) analysis to identify 28,000 eligible parcels covering 20,000 acres, of which one-third was planted with grass.[7] They established a goal of reducing "excess grass," encouraging people to use an acre for gardening and landscaping but to convert the rest to woodland. Baltimore County identified 7,100 acres of excess grass in rural residential lots alone. This is significant not only because wooded land better protects shared resources such as drinking water reservoirs, but also because it reduces pressure on farmers who are actively using their land. "If we were able to reforest the excess lawn in rural Baltimore County," Outen observed, "we could expand the forest cover in those areas by 19 percent."

To pursue this goal, Outen and his colleges conducted additional GIS analysis and reached out to residents. They found that "people are really eager for technical assistance—they get the message, they just don't know how to get there." The county intends to implement this approach on a wider scale to help meet their objective of finding an additional 1,500 acres of forest by 2025.

Another project the department has used to address the highly fragmented distribution of forestland in Baltimore County is called Multi-Owner Patch Base Management. The department is collaborating with the Maryland Department of Natural Resources and the Alliance for the Chesapeake Bay to identify high-functioning patches of forest and reach out to the owners of that land. They offer to conduct an assessment of the health of the land and work with the owners to develop management goals, if they wish. For example, they might find that the forest is denser than would be optimal for its forest type, or lacking in vertical canopy

---

[7]GIS is a method of amassing geographic data in digital form so that large volumes of data can be analyzed spatially.

structure. The team translates such findings into lay terms, provides ideas about how to move the forest to a healthier state, and guides owners to state and federal resources for cost sharing.

MINIMIZING THREATS TO FOREST HEALTH AND SUSTAINABILITY

Colorado's forests are not in very good condition, stated Lyle Laverty, a former assistant secretary of the interior for fish, wildlife, and parks and now a consulting forester based in Colorado. More than 4 million acres of lodgepole pine trees have been killed by the mountain pine beetle, and wildfires in the past decade have damaged large segments of forest. These developments are the direct result of long-term stress brought about by climate change, he said. Western ecosystems, Laverty explained, developed as fire-adaptive systems, but the appeal of living in the Colorado mountains has attracted large numbers of people over the past few decades. That has changed both the ways fires develop and the human response to them. Wildfires are very common in Colorado and have become the primary reason why landowners are interested in managing their land, Laverty noted. A 2012 fire in Lower North Fork, for example, burned 14,000 acres, resulted in three fatalities, and destroyed 27 homes. A fire in Waldo Canyon in 2014 burned more than 18,000 acres and 346 homes, resulting in $353 million in estimated property damage. Overall, the fire damage in 2012 covered 400,000 acres, releasing, Laverty estimated, approximately 10 million tons of carbon into the atmosphere.

Colorado is experiencing a lot of change, he noted, both in the demographics and in the values of those who live there, and in the forest conditions that shape their lives. Historical photographs of the Bitterroot National Forest in Montana and Idaho illustrate the degree of change that has also occurred in Colorado forests (see Figures 3-1 through 3-4). At present, the western portion of the United States faces persistent and pervasive fire risk, in part because changes in forest composition have meant high concentrations of flammable material in forests. A few decades ago, a 5,000-acre fire was considered major, Laverty pointed out, but today Colorado experiences fires covering 80,000 to 100,000 acres.

Responding to fire damage is very expensive, Laverty pointed out. The Denver Water Authority spent approximately $50 million, or $5,000 per acre, to clear sediment after a 1996 fire in Buffalo Creek. "We can do a lot of good forestry work for $5,000 an acre," he added.

Laverty explained how Colorado is working with landowners to address these problems. The challenges include establishing relationships with landowners and generating awareness of risk. In the Front Range region, for example, approximately two million people live in 800,000 homes that are at high risk from fire. He and his colleagues are trying to help these communities become more fire-adaptive so they are more resistant to damage and can recover more easily. Fires do provide "teachable moments," he pointed out. In the aftermath, the benefits of working to make one's land fire-adaptive can be very obvious if some neighbors have done so and others have not. A defensible space around a property is important, he noted, but communities that address landscape from this perspective are even better protected.

Laverty and his colleagues think about mitigating the impacts of forest fires in terms of five principles: communication, coordination among counties and agencies, cooperation, collaboration, and connection. Establishing connections is particularly important because many Colorado landowners reside outside of the state. "They may not even be aware that they are in a high-risk area," he noted.

Boulder County, which is very progressive politically, is one jurisdiction that has focused on getting the word out to owners about fire risk, Laverty explained. The county attaches an aerial photograph of a taxpayer's property to the tax bill to emphasize the need for "defensible space" around homes that will retard fire. The Colorado Forest Service has developed a web-based tool that landowners can use to identify the risk to their property. When owners key in the address of their land, information appears showing the risk, including potential flame length and speed. This tool is intended not only to increase awareness of risk, but also to provide information about mitigation and planning, and to motivate owners to take action.

The job of communicating with people about what can be done is critical, in Laverty's view. Fire suppression in the western United States is costing approximately $11 million per day, he concluded, and "we can do a lot more with that than just to fly a bunch of air tankers around." Fires are an opportunity not only to educate people about mitigating the risks, but also to show people how climate change is having significant impacts on their local areas and their lives.

CROSS-CUTTING IDEAS

In discussing the examples of communicating about forest management and climate change, several presenters noted the importance of thinking regionally, rather than locally. For example, Walkingstick noted, the invasive species Cogon grass is in neighboring states but not yet in Arkansas. "It is a huge problem that all of us are going to have to face," she commented. It is important to move beyond boundaries that may exist among agencies and universities and focus on the ecosystem and its needs. This is difficult, she acknowledged. Labich agreed, but suggested

FIGURE 3-1 Bitterroot National Forest, 1909.
SOURCE: U.S. Forest Service.

FIGURE 3-2 Bitterroot National Forest, 1948.
SOURCE: U.S. Forest Service.

FIGURE 3-3 Bitterroot National Forest, 1979.
SOURCE: U.S. Forest Service.

FIGURE 3-4 Bitterroot National Forest, 1989.
SOURCE: U.S. Forest Service.

that coordinating among partners that share a geographic region is critical. This may be easiest, he observed, if an organization can serve as the host and facilitate effective collaboration. Sagor added that "regional means different things in different contexts." In Minnesota, the extension system has moved away from a county-based system to one in which a statewide forestry team divides up areas of responsibility by content (e.g., agricultural forestry, forest economics and taxation, forest ecology and management, and invasive species). There are also increasing opportunities for regional programming at the watershed level, he added. Working across states is difficult, he noted, in part because local concerns are vital for engaging landowners.

Lyman explained that a range of factors has helped motivate communities in northern New England and eastern Canada to purchase and manage forestland, and changes in the forest products industry and land ownership have affected people on a personal level. Jobs have been eliminated and cultural traditions have been affected. For many communities, the possibility of determining for themselves what will happen with their land has been a major influence. Some come to the idea through a focus on managing growth, whereas others are focused on protecting a landmark. Economic concerns—such as promoting recreation and tourism or protecting logging operations—may also motivate them to manage and protect forestland.

In response to a question about efforts to reach landowners who are members of minority groups, Walkingstick noted that Arkansas's extension has focused on women landowners, but acknowledged that they need to do a much better job reaching out to minorities. Labich agreed, noting that though there has been very little focus on minorities in the New England region, his group bases its work on the understanding that "not all woodland owners are the same." (This theme is further explored in Chapter 4.)

Sagor added that Minnesota has a reasonably large American Indian community and that extension has worked with the Fond du Lac reservation and other groups to understand and respond to their needs. For example, he noted, he and his colleagues have recognized that the Native American communities are very strongly connected to the natural areas in which they live, but are not generally personal owners of forestland. The ash tree is a species with a very high cultural significance in the region, and he and his colleagues have looked for ways to take that into account in addressing the problem of the invasive emerald ash borer.

Finley concluded the session by noting that many of the presentations and discussions stressed the need to understand the values and goals of forestland owners. Thus, he stated, successful extension and education about climate change and forest management requires the tailoring of messages and programs to the needs and values of forestland owners.

4

Engaging Forestland Owners: Perspectives from the Social Sciences

Social and behavioral scientists have identified ways to communicate effectively about complex ideas and to change the way people behave, and have explored the many factors that influence behavior change. Four social and behavioral scientists discussed insights from that body of research and the implications for communication about climate change. A panel of social scientists and foresters drew on their experiences working with forestland owners in offering their responses to the presentations.

IDENTIFYING EFFECTIVE MESSAGES

Purnima Chawla, executive director of the Center for Nonprofit Strategies, linked ideas raised by the case studies discussed in Chapter 3 to concepts from the fields of social marketing, organizational behavior, and human decision making. She organized her observations around four elements that make communication effective. First, a lot of research documents the importance of the actual communication behavior, and which features make it easier or more difficult for people to understand one another. The content—the reason for communicating something in particular—is also important, she noted. So, too, is the context of the message: Its source, the place in which it was delivered, and "the contextual or emotional cues in the environment or in the communication materials all set the frame for the messaging," she explained. Last is the way people remember the message.

She elaborated on the relevance of just one of these—the content—in the context of communication with forestland owners. There are four basic types of content, or reasons why people want to use communication to induce others to do something, she explained. One is to convey a strategy for solving a problem. Another is to identify a valued outcome the listener can obtain. These two are easy messages to convey, she explained, and often fit naturally with the circumstances facing forestland owners—such as fire risk or the possibility of increasing the benefits of timbering or recreational use of forestland. The problem with both is that listeners have many issues to pay attention to, so uncertainty about the timeframe or urgency of the problem or benefit may undercut the message.

Uncertainty about the urgency and timeframe is an obvious problem with messages about climate change, Chawla explained. What may or may not happen in a particular place, when and how it may happen, and the likelihood that a particular action will make a difference, are all uncertain. Similarly, it may be that one day in the future carbon credits will yield profits, but when and how much is not known. The risk of wildfire, she noted, is more immediate than the risks of climate change. It is also true that landowners are often likely to be motivated to take immediate steps to protect their land from fire, whereas, for many, the fear of climate change may be overwhelming if they feel there is little they can do about it.

The two other types of messages may be more effective when it comes to climate change, she explained. Connecting behaviors to a value people hold is one effective way to communicate a message; a value that many forestland owners identify with is stewardship, Chawla observed. "People spend a lot of energy doing things that have no immediate positive outcome for them," she explained. "One motivation for doing so is to reinforce their impression of who they are and who they would like to be." The difficulty with this message, however, is that values and behaviors should be understood as a network, and a particular set of values may point to several different behaviors. Thus, the challenge for the communicator is to connect the recommended behavior to the value most likely to be a motivating influence.

In other words, she explained, "people may say 'yes I am a good steward, but I don't see this behavior as an expression of my stewardship ethic.'" In such cases, she added, repetition of the rationale can be very effective. For example, she observed that the connection between building a legacy and estate planning with stewardship of the forest itself may not be apparent immediately because "it is not a natural connection people make." But if the message is repeated, the target audience has more opportunities to consider it and recognize that stewardship can mean taking a longer perspective and planning for what will happen to land after it changes hands. "Usually attitudes and values are very broad," she added, so "you really have to spell it out and be convincing."

The fourth message strategy, Chawla explained, is social. Here the message is not that the landowner will take an action because of his or her sense of values and goals, but rather because of a desire to belong to the tribe of people who behave a certain way. The message might build on the idea that "people like me are doing this," or on beliefs about what others will think of particular behaviors. The logic of this sort of message suggests focusing more on positive things than negative ones, Chawla explained, citing an example from another field. Researchers have found that a message about how much damage tax evasion does to society was less effective than one that highlighted the fact that the majority of people actually do pay their taxes. The idea, she explained, is to "establish the norm."

There is no one right message strategy for any situation, Chawla observed, and "each has its pitfalls." Crafting an effective message, therefore, requires segmentation. "Not everybody is going to do things for the same reason, but you can confuse people enormously if you give eight reasons to do something instead of one," she added. Thus, it is necessary to address one target at a time, based on a clear understanding of what that group values.

Chawla closed with the observation that the relationship between attitudes and behavior is neither linear nor one-directional. "Sometimes," she explained, "if you can get people to do something, that actually strengthens their identity." At the same time, asking repeatedly that people take a particular action is more effective than asking them to do multiple things and hoping they will do one of them. Similarly, she concluded, a person who has agreed to one small thing is likely to be more receptive to a follow-up request to take a more significant action.

THE ROLE OF COMMUNITY

Maureen McDonough, professor of forest sociology and social forestry at Michigan State University, offered perspectives from the field of natural resources sociology about social relationships within communities, as they pertain to forestland owners. She began with approaches she observed in the case studies that align well with the research literature she tracks. First, she noted, sociologists consider that human beings exist as members of networks and communities, and so she was pleased to see excellent examples of network-building in the case examples. For example, identifying representative community members who can reach out to their neighbors and building communities of volunteers fit well with the sociological approach.

Trust is very important to communities and is the basis for people's willingness to engage in a new behavior, McDonough added. Researchers have found that just as trust is a key variable, so, too, lack of trust can be a significant obstacle. Thus, she noted, it is important to try to understand how much trust target audiences have in the messengers who are responsible for reaching them, whether they are neighbors, professionals, or government officials.

On the other hand, she noted that social science could be useful in ways she did not see reflected in the case examples. Most important would be to better understand the complex social contexts in which people make decisions. She noted that she and her husband used two cars to drive an identical commuting route for many years, even though they were committed conservationists, because other factors—such as the decision to drive children to sporting events—affected their lives. "People say they will do things," she commented, and "they may want to do them," but it is important to understand the obstacles that arise in their personal communities. "It can be futile to try and change people's values," she added in answer to a question. Values have multiple sources and they are different across generations. Younger generations, she added, are more receptive to messages about climate change than their elders—they may have similar values but express them differently. Trying to convince people that "you are right and they are not is not going to be effective," McDonough observed.

Political science research also offers insights about how people expect to participate in decision making, she added, and she did not hear much discussion in the case examples about including landowners in decision making. Research, she noted, suggests that "if you want people to behave in a certain way, you need to empower them." It is best to involve people early in the process, she added, using a collaborative, consensus-building process. "This requires some behavior change on the part of natural resource professionals," she added. She heard a lot of discussion about how to share information that people need, but little about the knowledge they already have and how that can be tapped into and built on.

The information context is also important to consider, McDonough observed. Sources of information have multiplied exponentially in the past few decades, and those who want to reach forestland owners must "navigate that." It is much more difficult to get people's attention than it used to be. More broadly, it is also important to consider cultural influences. A few of the case study presenters talked about people's reasons for owning their land, she noted, and referred to "logical" reasons for acting in certain ways. Scientists do not necessarily define logic in the same way other people do, she noted. Research has explored how people think about what they believe, she observed, and suggests that "belief systems are very powerful, and they are not generally based on facts." To engage effectively with landowners, she concluded, will require moving beyond logic and consider that "most people make decisions based on their hearts and their guts…and scientists make decisions based on their heads."

PRINCIPLES OF COMMUNICATION

Many of the practices described in the case examples correspond to theoretical models from the social sciences, Joe Heimlich, professor and extension specialist at Ohio State University, observed, but he said that a fuller understanding of those theories and how they function would make it possible to improve engagement with forestland owners. He focused on communication, noting that it may seem simple: "coming up with an idea, encoding that idea, and transmitting it through some kind of medium toward a target." People tend to expect that others receive their messages in the way they are intended to come across, he explained, but in reality communications are interpreted through filters that may significantly complicate the transmission. The basic question—of what goes wrong and keeps people from receiving messages as they were intended—underlies much of the research in educational and learning psychology, as well as communications, he added.

There are many distinct factors that distort communications, Heimlich noted, that have generated lines of research. Noise, or distortions that affect the encoding and decoding of the message, is one (Goldstein, 2014). Another is perception (Berliner, 1955), the process by which people use their senses to collect information and organize and filter it. "How we perceive things affects how they are framed in our minds," Heimlich explained. Similarly, people have expectations that shape the way they perceive communications. "If I expect a conversation is going to be hard, it will be hard," he noted, but "if I expect a person will give me good information, I will receive it."

Information load is another potential obstacle (Lavie et al., 2004). There are many influences and stimuli in people's lives, Heimlich explained, that may get in the way of their capacity to hear or receive a given communication. He noted that gambling casinos take advantage of this phenomenon by making their environments so loaded with sensory input that visitors' senses are numbed, and "because of that, many people do exactly what they [casino operators] want us to do when we are in those places." Limits on people's skill at listening are yet another potential obstacle to communication, Heimlich added. It is "an art that we tend not to practice very much—we hear but we don't listen, we don't really get the nuance." Context is key to listening with understanding, he added, noting this as one reason why the tone of e-mail messages is often puzzling.

These insights can be used for good or evil, Heimlich observed. To use them in engaging constructively with forestland owners, it will be helpful to get the right stimulus to the target audience in such a way that the audience responds in the desired way. A starting point is to explore landowners' habits and behaviors. Much of what people do on a daily basis is embedded in routine, he noted. For example, most of the actions a person takes when getting into the car and driving to work are routine, but if one took time to think about each activity associated with driving to work "it could take 10 or 15 minutes to get out of the driveway." This is important, he explained, because it means that to get a person to change the way he or she behaves, it is necessary to look at a whole pattern of routine behavior.

Thus, Heimlich went on, "telling people to do something is not teaching them a behavior," while to teach the behavior is to help the individual understand "how to practice, apply, and perfect the skill in the context of his or her routines." This would mean "starting with the behavior you want them to take, working backwards to see where would you embed that in their practice," and then determining how the new behavior might fit the existing routine. He acknowledged, in response to a question, that behavioral theories identify different points as the optimal levers for getting a person to change a behavior.

Heimlich noted that it is important to recognize, as many of the presenters had done, that people have a range of reasons and motivations for the things they do, and that they themselves may not always recognize all of the factors that influence their actions. They also may not receive or respond to messages that are not present when they are needed. That is, those working with landowners may believe they have done all they can if they "put things in newspapers, sent e-mail blasts and broadcasts, talked with everybody" but people still say they had not heard about the message. The reason, Heimlich explained, is that the messages were "part of the noise that they were not paying attention to."

Sociocultural theory,[1] Heimlich said, demonstrates the importance of culture in communication. Social role is a significant determinant of people's actions, and it is a dynamic, rather than a static, concept. For example, a message to forestland owners about the land as a family legacy may come across differently depending on whether the recipient is hearing it as "the child of his grandparents or as the parent of his children and grandchildren."

---

[1]Sociocultural theory is a psychological theory that stresses the importance of cultural contribution to an individual's cognitive and social development.

What all of these insights point to, in Heimlich's view, is that "we can't change people, we can change how we frame things." Specifically, he concluded, this would mean that those who want owners to take an action or change a behavior need to try to understand how that change would relate to existing routines, the context for the shift. Other questions to consider include: What is the potential return on investment for the owner who might make a change? Does the landowner have the power and the resources to make the change? Would the outcome clearly be at least equivalent in benefit to the current approach? What are the trade-offs? Are there intangible benefits that can be made clearer?

COMMUNICATION ABOUT CLIMATE CHANGE

Many social scientists have focused specifically on communication about climate change. Christopher Clark, assistant professor in the Department of Communication at George Mason University, described some of that work and how it might relate to engagement with forest-land owners. Public opinion about climate change is surveyed twice a year in the United States[2] and the survey data have helped him and his colleagues consider strategies with respect to three goals for engaging people on the topic. One is to engage people on a cognitive level—so that they will understand the scientific consensus that climate change is taking place. People have emotional reactions to that knowledge, Clark added, so it is also important to engage them on an affective level, whether their reaction is one of optimism, fear, worry, or something else. The third goal is to persuade people to change their behavior, both personally—such as driving more fuel-efficient cars—and at a macro level, by supporting policy changes, for example.

"I've never come across an issue that's as politically polarizing, ideologically, left/right, red/blue, democratic/republican, as climate change," Clark observed. Several lines of research hold promise for meeting the three engagement goals even in this polarized context, however. As many others at the workshop had noted, the basic strategy is to understand the audience, design messages based on that, and use trusted sources to deliver it. Building on that, he explained, begins with the profile of six distinct groups of Americans' understanding of climate change, shown in Figure 4-1 (also discussed by Geoffrey Feinberg as the "Six Americas"; see Chapter 2).

Clark and his colleagues have tracked changes in the opinions of these groups over time. The most recent data show stark differences in

[2]See http://environment.yale.edu/climate-communication/article/extreme-weather-public-opinion-September-2012 [February 2014].

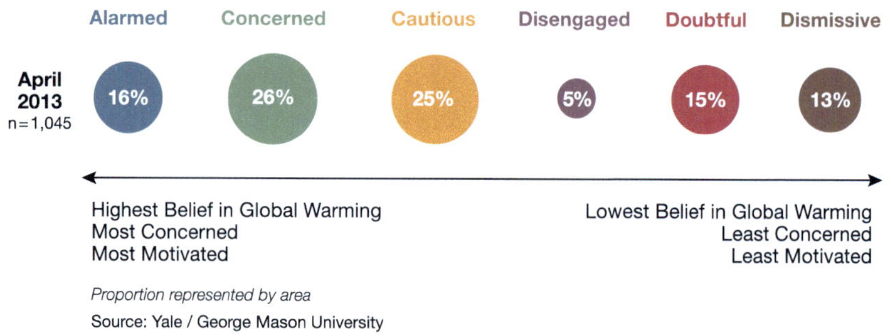

FIGURE 4-1 Global warming's "Six Americas."
SOURCE: Leiserowitz et al. (2012b).

thinking among the groups, as Figures 4-2 and 4-3 show, and also differences in how important these groups believe the issue is (Figure 4-4). These results clearly show polarization, Clark noted.

Clark and his colleagues have also tested messages with the six identified subsegments of the U.S. population. This research suggests that

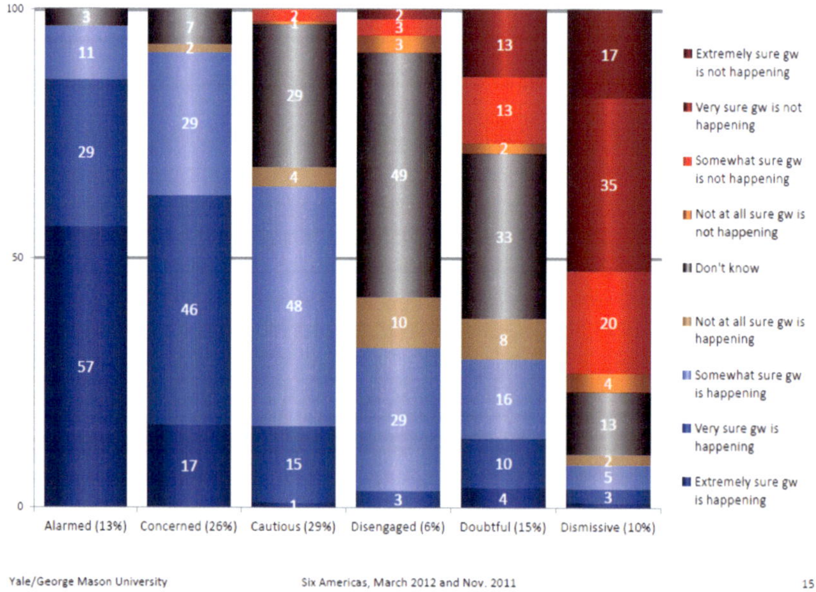

FIGURE 4-2 Certainty of belief about the reality of global warming, March 2012.
SOURCE: Leiserowitz et al. (2012b).

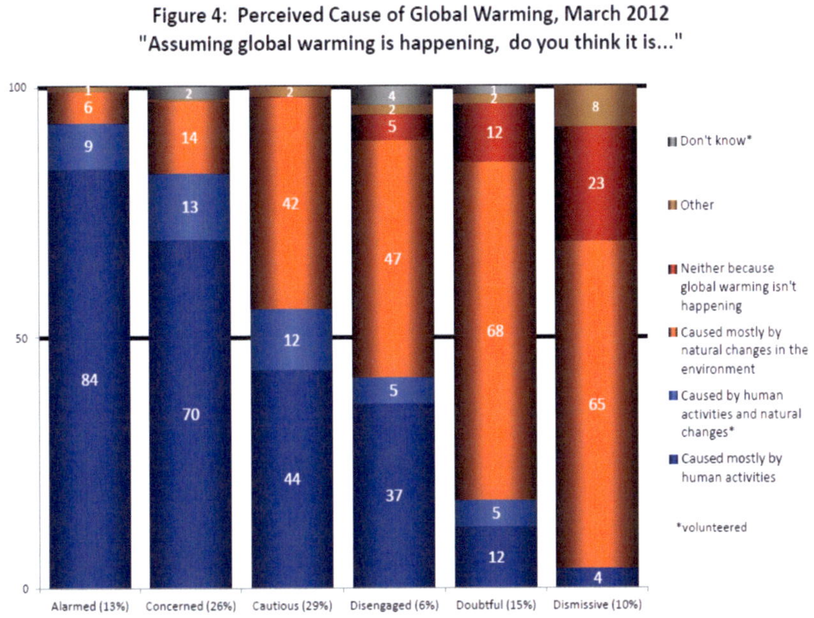

SOURCE: Leiserowitz et al. (2012b).

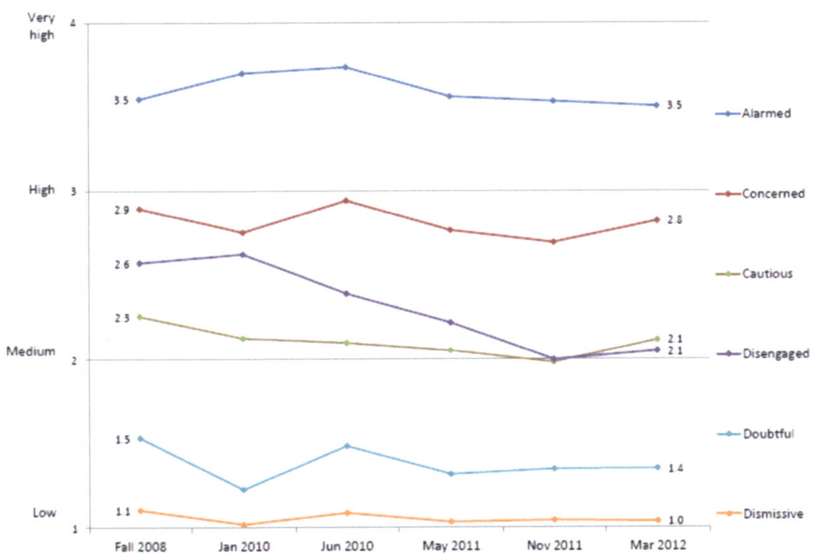

FIGURE 4-4 Priority of global warming for the President and Congress, 2008-2012. SOURCE: Leiserowitz et al. (2012b).

perceiving that a consensus exists among scientists that global warming is happening is "a gateway belief that can drive other beliefs" about how likely it is that climate change could harm an individual or community, or the merits of proposed policies. Clark noted that the Consensus Project is one group that has focused on spreading understanding that 97 percent of climate scientists agree that climate change is caused by human actions.[3] However, Clark noted, the six groups vary significantly in their responses to a simple visual image making that point.

He and his colleagues have also found variation in people's responses to different ways of framing the concept of climate change. If people are asked how worried they are about it, half to two-thirds say they are worried, he explained. However, the majority do not see it as an urgent threat: 35 percent believe global warming will cause a moderate amount or a great deal of harm to them personally, while 63 percent believe it will cause harm to future generations or plants and animals. In other words, the further removed the threat seems to be "from me and mine," Clark explained, the less concerned people are about it. He added, in answer to a question, that people's responses to personal experiences with events that could be evidence that climate change is happening, such as extreme weather or drought, vary across the Six Americas. "If I'm in that dismissive group," he noted, it may be that "no amount of odd weather could change that." For those in the middle, however, extreme weather may be a "teachable moment," he added. However, because the extremity of weather is relative, it is not possible to know how strong a storm would have been were climate change not happening, or how long and severe a particular drought would have been without climate change.[4]

Climate change as an issue can be framed in a variety of ways, and although it may seem "cavalier, " he commented, to develop "interpretive packages" that emphasize particular facets of the issue, he suggested this may be the most effective way to persuade people to accept that action is needed. For example, some evidence suggests that discussion of risks to public health posed by climate change, and the actions that might mitigate those risks, may be more compelling to audiences who have resisted other approaches to the topic. Thus, discussion of heat waves, extreme weather, infectious diseases, or improving air quality may be effective. Discussion of economic damage that will come with the effects of climate change, such as the losses associated with Hurricane Sandy, is another approach that has been effective. Still, Clark acknowledged, the "doubtful" and

[3]See http://theconsensusproject.com/ [January 2014].

[4]For further discussions on the significance of the Six Americas audience segmentation on climate communication in general, see National Research Council (2011).

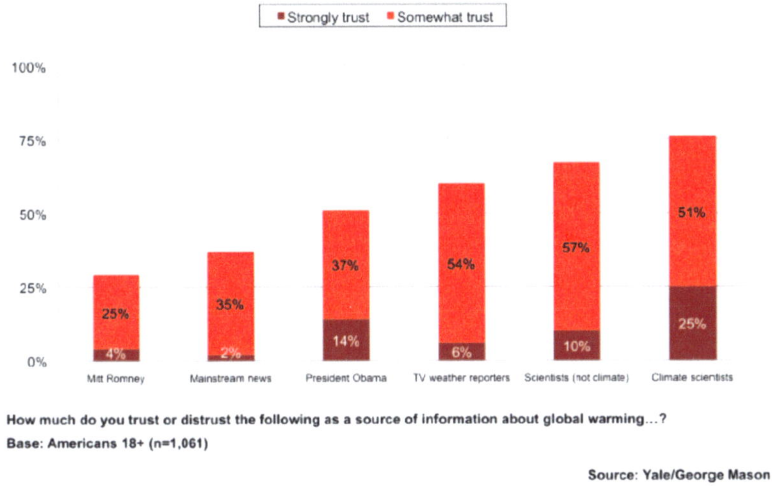

FIGURE 4-5 Public trust or distrust of sources of information on global warming, March 2012.
SOURCE: Leiserowitz et al. (2012).

"dismissive" groups are less responsive even to these messages than the other groups are.

Whatever the message, Clark went on, it will be most effective if it comes from a source the recipient trusts. Figure 4-5 shows data about how much trust people have in different sources of information about climate change. "People tend to trust climate scientists," Clark pointed out, "but they don't really know who they are." Since people also trust weather reporters, Clark and others have begun to work with these reporters to encourage them to incorporate factual messages about climate change into their forecasts.[5] This would mean not linking a particular weather event directly to climate change, he added, but elaborating on the context to help people experiencing a heat wave understand, for example, that heat waves are becoming more common because of larger patterns.

Climate change communication is a matter of "engaging hearts and minds," Clark concluded, or cognitive, affective, and behavioral engagement. He believes the analysis of the Six Americas can be applied in any context in identifying messages that fit specific goals. A partici-

[5]Clark referred participants to a segment about the role weather reporters can play that was aired on National Public Radio in February 2013. See http://www.npr.org/2013/02/19/171832641/forecasting-climate-with-a-chance-of-backlash [January 2014].

pant observed in response that forestland owners already loosely group themselves by affiliating with various organizations, including the Forest Industry Trade Association, American Forest and Paper Association, state forestry agencies and extension programs, and American Tree Farm System, as well as land trust and community forest associations. These organizations may have very different objectives, which include protecting and promoting financial investment, sustainable stewardship, and other goals. Some focus on the interests of individual landowners while others look at "the whole enchilada: the air, the water, the wildlife habitat," he commented.

PRACTITIONERS' RESPONSES

Claire Layman, an extension public policy education specialist at Michigan State University (MSU), spoke about her experience working with the MSU state extension program. She was astounded when she began, she commented, to find that extension was largely not addressing climate change. She identified others within extension who were eager to share the information, and they began with field crop farmers. From talking with this group informally and in focus groups, she and her colleagues found that many believed they were being blamed for climate change, that they were not seen as stewards of the environment, and that they would be paying a regulatory cost. However, many also appreciated the opportunity to be heard, and Layman found that the conversations and information gathering were the start of a productive relationship.

Today, she and her colleagues are working with two communities in Michigan to create climate change adaptation plans. In one of the communities, she noted, a transportation planning organization advertised a public meeting to collect public input for their long-term planning effort that cited a desire to hear people's thoughts about climate change. "This was like a siren call for the alarmed and the dismissive," Layman explained. The planners wanted to avoid a scenario in which a scientist would make a presentation and then be "shot down" so they began the meeting by asking participants to suggest ways they thought the climate might be changing. Nevertheless, she added, the reactions were predictable and the "dismissive crowd was angry that we presupposed that the climate was changing." For her, the lesson was that "even though extension is a trusted source, we came in from outside that community, we partnered with a source that many people saw as a bloated government entity, and we talked about climate change." In retrospect, she believes that the approach they took with the farmers was more productive, and that if the topic of climate change is divisive, working around it may be wise.

Victor Harris, publisher and editor of *Minority Landowner Magazine*, offered the perspective of minority landowners. He and his colleagues, he explained, have the objective of providing valuable information about land management to their readers. They have learned, however, that it is important to begin by working to understand what the forestland owners' objectives are, and provide guidance that helps with that, rather than with objectives of their own. He added that they try hard to convey that "every time you put in a crop, however many times a year, every year for how many years—whether it is corn or soybeans or tomatoes or trees—every time you touch that crop, you're doing research. Your actions or inactions, that's research."

It is important to understand, he explained, that landowners understand a great deal about their land. Scientists have a language they are accustomed to using, he added, but in talking to forestland owners, "you have to translate that into something they understand." Thus, for Harris, two points in the presentations were most striking. First, because the language used will determine how the message is received, it is vital to "deconstruct the message before sending." Second, a critical point was that effective messages come from trusted sources.

James Houser, owner and consulting forester for James Houser Consulting Foresters, LLC, began by noting that as a professional forester who supports landowners in managing their property, he has noticed significant skepticism in his region (Texas) about politicians and the scientists who advise them. The majority of the landowners with whom he works do not live on their land and most use it primarily as an investment. They may have inherited it and are concerned about taxes and uncertainty—many are looking for information to help them in deciding whether to keep the land and develop timber growth or simply sell and take the profit. They want to know, Houser explained, "What's my rate of return over the next years? Why shouldn't I just clearcut it, take that cash now, let it grow back on its own, and do that about every 20 to 25 years?" A timber forest in Texas can be very profitable, Houser added, and most owners need to have a reasonable rate of return on investment to keep the property, even if they want to use it for recreational purposes as well.

Consulting foresters, who work only with non-industrial, private owners, view education as a key aspect of their mission. He and his colleagues need help, he explained. The forestland owners who are their clients "do not want to read scientific papers, they want to know the bottom line." They are very concerned about the danger of fire, and Houser said he appreciates the link from that concern to climate change. They generally view their landownership as a business and are concerned about competing with larger, industrial landowners. They are interested in sound, professional practices for logging, but he said for the most part they are conservative Republicans who are not interested in environmental messages, which they view as liberal. Thus, Houser and his colleagues tend not to make the connections between good management and benefiting the climate.

Houser's approach in working with owners is first to establish their own goals for their land and then help them to understand some principles of forest management. For example, he recommends strategic cutting to allow the forest to flourish, and he argues that this is a way to improve the rate of return on the investment, rather than on capturing more carbon through improved growth. It is important for his trustworthiness, he added, that he himself is a forest owner who manages his own land in ways he is recommending to them.

Alton Perry, speaking from the perspective of the North Carolina Forest Service, agreed with the previous comments. North Carolina conducted a forestland assessment in 2007 that examined all aspects of the state's forest resources as well as the risks posed by climate change. He and other extension workers used the information for education programs. For example, they realized that they had been advocating that trees be planted 622 to the acre, but the result was too dense. Widening the spacing allows for greater plant and wildlife diversity, they found. Like Harris and Houser, he added, he and his colleagues provide information of this sort as part of an effort to support owners in pursuing their own management goals. "As foresters and forest technicians," he added, "we try to identify sound forest management to help them reach those goals."

He and his colleagues have also found that owners are wary of government officials who may offer a message but do not stay around to help the owner implement a response. Extension workers gain trust when they are responsive to landowners' questions, he added. In a climate of limited resources, they have found other agency partners to help support and deliver messages to the landowners. For example, they are working with the Roanoke Electric Cooperative to help guide local forestland owners in addressing forestland management issues.

Amanda Mahaffey, northeast region director for the Forest Guild, described the experience that members of the Forest Guild[6] had in working with woodland owners. She agreed with earlier presenters in believing that climate change outreach depends on the contributions of multiple actors. She sees a lot of value in bringing together diverse groups, as the

---

[6]The Forest Guild is an organization of foresters and associated resource professionals who practice and promote sustainable and ecologically based forestry to benefit the forest and the human communities that depend on it. See http://www.forestguild.org/About-Us.html [December 2013].

Roanoke Electric Cooperative has done, to develop a "shared understanding of what can be done for everyone's mutual benefit."

Foresters are not generally trained in social communication, media studies, and marketing, Mahaffey noted. Foresters tend to choose their career path because they prefer being alone in the woods to public discussion, she added, but "we have to learn." Foresters who understand the role of climate change have a responsibility to be "that trusted resource" for other landowners, in Mahaffey's view. They need the communication tools and strategies that are being discussed at the workshop to meet that objective. Developing the communication skills of the foresters who can spread the word is a goal separate from the goal of reaching new audiences. "We need the psychology and sociology 101 course," she noted.

Karl Dalla-Rosa, forest stewardship program manager, closed with the perspective of the U.S. Forest Service, highlighting two key aspects of the forest stewardship program. One is communicating to the general public why private forestland is important and the role its owners play in adapting to and mitigating climate change. Recognizing how controversial the issue of climate change remains with some people, though, he believes it may not be the ideal platform for getting out core messages about forest management. The second aspect is engaging the forestland owners themselves. He noted that the workshop presentations seemed to support the direction the forest stewardship program has taken, which the Forest Service calls the landscape stewardship approach.

The agency's objective is to work with owners proactively to encourage them to take actions that benefit the climate and also support their own objectives for their land. The Forest Service works through existing social structures and communities, he added, to develop an appreciation for the entire landscape, beyond the boundaries of an individual's property. They have sought out peer-to-peer networks, hoping to leverage the expertise and knowledge landowners already have, following up with professional advice where it is needed. Nothing they would recommend that landowners do, he added, would be "in conflict with what a landowner would do to manage a property for forest health or resilience or those things."

Dalla-Rosa stated, however, that eventually it is the Forest Service's responsibility to "relate what we're doing with that landowner to broader climate change objectives." Ideally, the service would engage landowners not only in managing their own land responsibly, but also in spreading the word to other landowners about what they are doing and how it relates to climate change.

5

Recurring Themes and Questions

The workshop allowed a significant amount of time for structured and informal discussion, some in small groups. Several themes and questions emerged repeatedly.

CONSIDERING THE MESSAGE ITSELF

The workshop was based on the premise that there are clear messages that professional foresters would like to share with landowners, but presenters and participants raised questions about the content of those messages. One noted that it would be important to be clear about "the underlying facts that we all agree on" with respect to what landowners ought to be doing. Many cited the primary goal of keeping forests as forests, noting that idea has broad appeal. However, some suggested that more work needs to be done to elaborate these messages.

"How do we take all of this education and all of this science and water it down to a sixth- or eighth-grade education level—how do we take all of this great information and bring it down to the level where people can understand it, accept it, and take action with it?" asked one participant. "We are potentially putting together a master strategy for how to communicate about climate change to family forest owners, individual owners, community owners, but the missing piece is: What do we actually tell the landowners to do differently than they're doing now," another commented. "If we can identify those actions, those critical steps, those basic steps that landowners can take and train our extension specialists and our foresters to convey those actions as a step, then that will be a great outcome," another noted.

Several participants observed that simply raising awareness of the issues of forest management and the idea that it will be beneficial to prepare now for coming changes will be valuable. "Persuading someone that they really should care about this is a huge undertaking," one person commented. "We don't know. Nobody knows, climatologists don't know, forest ecologists don't know, we don't know what the future is going to hold exactly," another pointed out. But, the participant continued, "We do know there are certain attributes that certain ecosystems have and some don't that predispose them to higher or lower levels of risk based on outcomes of a variety of different future conditions." Many seemed to agree that the best focus would be on very practical implementation of practices that will be beneficial in the short term while also beneficial from a climate perspective.

Another audience member suggested that the focus should not be so much on what to tell landowners, but on preparing the service providers to have the best information and science available so they can integrate that information into guidance tailored to the specific forest ecosystems and threats in a particular region. Another agreed, noting that the service providers are "all over the place." University extension and state agency staff are likely to be well informed, this person added, but in some regions there are foresters "who don't need to be licensed and probably have no continuing education requirements, and maybe some of them have never tried to update their knowledge about how you manage forests under these conditions. And we have the whole spectrum across the country."

On the other hand, another noted, "we may not know exactly where we are headed, but we have some good messages: Take care of your forest, keep it healthy, engage professionals, talk to your neighbors, learn from others. What we have to remember is that people care about trees and forests."

IS IT BETTER NOT TO TALK ABOUT CLIMATE CHANGE?

A related point made by several attendees was that in many cases, it might make sense not to speak directly about climate change. They noted that doing so "puts a lot of people off" but also made the point that there is an ethical responsibility to share this knowledge. The reality of climate change also "ties many different threads together, especially in terms of mitigation," one person noted. It may be a tougher sell to argue for forest-related policies without a larger narrative of climate change benefits and risks. One participant suggested that discussions about the impact of climate change on forests and the role forest management can play in addressing climate change are needed. Such discussions can provide a basis for tackling the regulatory environment, but are also essential in preparing a broader social consensus on climate change and its long-term implications. While it might be prudent to leave climate change out of any one forest management discussion, in order to be more effective with a skeptical audience, the issue itself cannot be avoided altogether, the participant said.

LISTENING TO LANDOWNERS

Discussion returned frequently to the importance of engaging landowners, understanding their own goals for their land, and getting a sense of what they already know and what their needs are before pushing recommendations on them. Many also highlighted the importance of working within the communities in which the owners live, rather than expecting to drop in and offer advice that will be respected.

One participant noted that he intended to go back to his community and ask landowners about what they are seeing in their woods—to begin a conversation through people's first-hand experiences. The conversation "doesn't have to be profound," he added; "it could be invasive species, it could be that this tree is leafing out or that tree is dropping its leaves or this invasive showed up or whatever." The value would be in encouraging landowners to communicate with one another so they can realize for themselves, as the participant noted, "'I think I've seen that too but I didn't know what it was,' or 'wow, it's not just me but it's happening all around.'"

Beginning with people who are already engaging with forest management and climate change issues is a promising way to engage in a community, many noted. If "you get your foot in the door" with a local group on other sorts of conservation efforts, for example, then later it might be easier to engage them in conversations related to climate change adaptation activities, a participant observed. "Whether it is legacy planning, community forest, peer-to-peer networks, targeted outreach groups, social media, stewardship forest, and keystone cooperatives, regional conservation partnerships extension, faith communities, there are audiences that we can get our foot in the door or have our foot in the door already," another added.

Another participant wondered if it would make sense to work only with those people who understand that climate change is occurring and would like information about what to do. "They already want to take action," this person observed. "As an educator it's a lot more fun for me to work with people who are motivated and interested and learning, and they're more likely to take action." It may be that once those groups have had success in implementing forest management, their stories will reach others. "We can point to real people on the ground, not just public lands where nice things have been done. It gives us stories that are real and based in those communities," he added. Several participants emphasized the power of stories, advising that "when you're reaching out to landowners, tell stories, not just the facts—you tell them convincing anecdotes, because everybody has facts."

ORIENTING MESSAGES TO THE OWNERS' GOALS

"You need to meet landowners where they are," many participants observed. "If you engage with a landowner, nine times out of ten it's at their invitation," one noted, adding that "whatever they're interested in doing, whether it's watching wildlife, whether it's hunting, almost in every case the things we would do for them as service providers or technical assistance specialists or natural resource professionals in some way, shape, or form is going to have a positive outcome for carbon, for climate." Various participants noted the potential for conflict when the goals and objectives that forestland owners and those who want to guide forest management have diverged. They emphasized how important it is that service providers utilize the whole range of tools education and social science research can suggest to align agendas for the long-term benefit of owners, society, and the forest ecosystems themselves.

"We tend to teach people in the way that we learned ourselves," another observed, but that may not be the best way to reach people. "Adults want to learn a specific thing," this person added. "If there is something they need, they go find information." If landowners have the opportunity to help identify what will work for them, they will be more open to it. "As they try it, they learn different things," one person noted, "and as they do that they are in a better position in the community to share it with somebody else. Peer-to-peer education is a powerful tool. There aren't enough of us to do the work that needs to be done."

It is critical to understand what drives landowners, many participants said. One pointed out that landowners need to get a financial return on their property or they will not be able to pay their property taxes: "You get basically three years and you've lost your land. People don't realize that driving force to raise money from our land."

Another noted, however, that "forest and people are all part of a system. We're all in this together. Some people are highly motivated by dollars, but people are motivated by other things." It's necessary to identify what those things are, he went on. "We heard about the southern forest and the northern forest and the eastern forest, and we heard about the woods in the backyard, all these things represent different places and different people."

On this point, another noted that if the goal is to protect forestland, there are tools. Easement programs and other policies can make a difference, he noted, but he suggested that "the best way we can keep forests as forests is to have a healthy and viable and robust forest product industry." If there are no markets, forest management is difficult. Without funds, conservation is difficult, he noted. "Without the private sector providing incentives, along with the public sector, we're never going to have enough public resources to ease or cost-share our way or give tax breaks, and we're going to need that private sector incentive. So without forestry infrastructure and without that private market, we could talk about management all day long, but who's going to do it? Are we going to be paying people to do it, or are we going to have people pay us to help us get the job done?"

EARNING TRUST

A related theme was the importance of earning the trust of the owners one wishes to reach. In some areas there may be a legacy of mistrust, as well as lack of clarity about the motives of different people and groups who offer advice to landowners, several participants noted. One commented that a "landowner may be thinking 'I don't trust you because you stole my grandpa's property, you even stole the land itself, you didn't give him what the timber was worth, why should I trust you?'" A participant from a tribal community agreed that this is important, noting that "we have become experienced with the history of people from the outside coming and telling us what's best for us, without having our voice heard in that process, and that creates a barrier right there."

Others noted that any attempt to change behavior can be met with distrust by those who may feel manipulated. A language of trust that is based on mutual respect and appreciation might be at odds with strategies for encouraging individuals or groups to behave or act in ways others prefer, as one participant noted; the challenge lies in establishing bridges between different interests and goals in ways that feel genuine and honest.

COLLABORATION

A participant noted that "getting the scientists and the practitioners in the room together is the only way to do this….The goals for preserving, protecting, and increasing forestland cannot be accomplished by foresters, forest ecologists, forest entomologists, forest economists, forest anything," he suggested. "There are not enough of us to do that, so how do we get that done? We need to rely on peers. And not just because we need more people to do it but because it's the right way to do it."

FINAL REFLECTIONS

According to the statement of task, a primary goal for the workshop was to explore effective strategies that service providers can use to engage forestland owners in discussions about climate change. The social scientists and practitioners who made presentations and participants offered many perspectives during productive discussions. A final question that emerged was how service providers might get the foundational education and training to professionalize their approaches in communication, outreach, consulting, and dialogue. Discussion highlighted the importance of better preparing future foresters for these challenges through post-secondary education and in other ways during their university education. Several participants also suggested embedding appropriate elements into a system of professional development and learning for the intermediaries between the science of climate change and the application through sound forestry practices across the nation.

References

Berliner, A. (1955). The Rorschach determinant in terms of visual psychology. *Optometric Weekly, 46,* 13-20.

Butler, B. (2006). *Family Forest Owners of the United States. A Technical Document Supporting the Forest Service 2010 RPA Assessment.* Newtown Square, PA: U.S. Department of Agriculture, Forest Service, Northern Research Station. Available: http://www.nrs.fs.fed.us/pubs/gtr/gtr_nrs27.pdf [May 2014].

Goldstein, E.B. (2014). *Sensation and Perception, Ninth Edition.* Belmont, CA: Wadsworth.

Gootee, R.S., Blatner, K.A., Baumgartner, D.M., Carroll, M.S., and Weber, E.P. (2010). Choosing what to believe about forests: Differences between professional and non-professional evaluative criteria. *Small-Scale Forestry, 9,* 137-152.

Grotta, A.T., Creighton, J.H., Schnepf, C., and Kantor, S. (2013). Family forest owners and climate change: Understanding, attitudes, and educational needs. *Journal of Forestry, 111*(20), 87-93.

Jones, S.B., Luloff, A.E., and Finley, J.C. (1995). Another look at NIPFs: Facing our myths. *Journal of Forestry, 93*(1), 41-44.

Lavie, N., Aleksandra, H., and de Fockert, J.W. (2004). Load theory of selective attention and cognitive control. *Journal of Psychology: General, 133*(3), 339-354.

Leiserowitz, A., Maibach, E., Roser-Renouf, C., Feinberg, G., and Howe, P. (2012a). *Extreme Weather and Climate Change in the American Mind.* Yale University and George Mason University. New Haven, CT: Yale Project on Climate Change Communication. Available: http://environment.yale.edu/climate-communication/article/extreme-weather-public-opinion-September-2012 [February 2014].

Leiserowitz, A., Maibach, E., Roser-Renouf, C., and Hmielowski, J. (2012b). *Global Warming's Six Americas in March 2012 and November 2011.* Yale University and George Mason University. New Haven, CT: Yale Project on Climate Change Communication. Available: http://environment.yale.edu/climate-communication/files/Six-Americas-March-2012.pdf [February 2014].

Leiserowitz, A., Maibach, E., Roser-Renouf, C., Feinberg, G., Rosenthal, S., and Marlon, J. (2014). *Climate Change in the American Mind: Americans' Global Warming Beliefs and Attitudes in November, 2013.* Yale University and George Mason University. New Haven, CT: Yale Project on Climate Change.

National Research Council. (2011). *Climate Change Education: Goals, Audiences, and Strategies: A Workshop Summary.* S. Forrest and M.A.Feder, Rapporteurs. Board on Science Education, Division of Behavioral and Social Sciences and Education. Washington, DC: The National Academies Press.

National Research Council. (2012). *Climate Change Education in Formal Settings, K-14: A Workshop Summary.* A. Beatty, Rapporteur. Steering Committee on Climate Change Education in Formal Settings, K-14. Board on Science Education, Division of Behavioral and Social Sciences and Education. Washington, DC: The National Academes Press.

U.S. Department of Agriculture. (December, 2012). *Effects of Climatic Variability and Change on Forest Ecosystems: A Comprehensive Science Synthesis for the U.S. Forest Sector.* J. Vose, D. Peterson, and T. Patel-Weynand, Eds. Portland, OR: U.S. Department of Agriculture. Available: http://www.fs.fed.us/pnw/pubs/pnw_gtr870/pnw_gtr870.pdf [January 2014].

Vose, J.M., Peterson, D.L., Patel-Weynand, T. (2012). *Effects of Climate Variability and Change on Forest Ecosystems: A Comprehensive Science Synthesis for the U.S. Forest Sector.* Gen. Tech. Rep. PNW-GTR-870. Portland, OR: U.S. Deparment of Agriculture, Forest Service, Pacific Northwest Research Station.

Walthall, C.L., Hatfield, J., Backlund, P., Lengnick, L., Marshall, E., Walsh, M., Adkins, S., Aillery, M., Ainsworth, E.A., Ammann, C., Anderson, C.J., Bartomeus, I., Baumgard, L.H., Booker, F., Bradley, B., Blumenthal, D.M., Bunce, J., Burkey, K., Dabney, S.M., Delgado, J.A., Dukes, J., Funk, A., Garrett, K., Glenn, M., Grantz, D.A., Goodrich, D., Hu, S., Izaurralde, R.C., Jones, R.A.C., Kim, S.-H., Leaky, A.D.B., Lewers, K., Mader, T.L., McClung, A., Morgan, J., Muth, D.J., Nearing, M., Oosterhuis, D.M., Ort, D., Parmesan, C., Pettigrew, W.T., Polley, W., Rader, R., Rice, C., Rivington, M., Rosskopf, E., Salas, W.A., Sollenberger, L.E., Srygley, R., Stöckle, C., Takle, E.S., Timlin, D., White, J.W., Winfree, R., Wright-Morton, L., and Ziska, L.H. (2012). *Climate Change and Agriculture in the United States: Effects and Adaptation.* USDA Technical Bulletin 1935. Washington, DC. Available: http://www.usda.gov/oce/climate_change/effects_2012/CC%20and%20Agriculture%20Report%20(02-04-2013)b.pdf [May 2014].

Appendix A

The Climate Change Education Roundtable

Climate Change Education (CCE) is becoming a major investment area for the federal government, and will involve among others, the National Science Foundation (NSF), the National Oceanic and Atmospheric Administration (NOAA), the National Aeronautics and Space Administration (NASA), the Environmental Protection Agency (EPA), the U.S. Geological Survey (USGS), the Department of Education, and the Department of Energy. CCE is defined as an effort that seeks to ensure that individuals and communities understand the essential principles of Earth's climate system and the impacts of climate change, and are able to make informed and responsible decisions with regard to actions that may affect climate and adaptation to climate change. Such broad goals for CCE demand a transdisciplinary approach that blends education, learning, social, behavioral, economic, and global Earth system science and requires careful consideration of related research evidence from each of these disciplines.

The National Research Council established the Roundtable on Climate Change Education in September 2009 to foster ongoing discussion of the challenges to and strategies for improving public understanding of climate science and climate change among federal agencies, regional and local government units, the business community, nonprofit, and academic sectors. Through its five meetings and two associated workshops over a course of approximately three years, the 30 Roundtable members were learning from each other and invited experts about issues such as the challenges inherent to climate change education; strategic approaches to designing interlocking programs in research, learning, workforce development, and public literacy; and strategies for accelerating the education of U.S. citizens about climate and climate change. The Roundtable brings together federal and state policy makers, educators, scientists, and communications and media experts. It includes a number of *ex officio* members from federal agencies with dedicated interests in climate change education, including officials from NSF, NOAA, NASA, USGS, the Department of Energy, the Department of Education, the Department of Agriculture, and EPA.

The CCE Roundtable provides an opportunity to bring together overlapping and complementary expertise from academic and professional disciplines that commonly do not intersect when addressing CCE. It also provides federal agencies with important foundational knowledge related to key aspects of CCE and learning, such as the nature and scope of existing efforts, achievable and measurable goals, challenges and opportunities inherent in developing a national-level CCE initiative, and areas where investments may provide the greatest leverage. Roundtable discussions also provide useful new insights for a variety of other stakeholders. The CCE Roundtable also provides a formal mechanism to support continued collaboration and cooperation across federal agencies on major future climate change education or other science education initiatives. Through Roundtable discussions, the work of the federal agencies can be coordinated with stakeholders from private and nonprofit sectors such that their efforts can be built to complement and enhance federal initiatives. Through its public workshops and published workshop summaries, the CCE Roundtable will also become a source for evidence-based information related to climate change education.

So far, the CCE Roundtable has sponsored four associated workshops. The first one focused on fundamental goals and objectives of climate change education and outreach to various target audiences, and ways to understand whether these goals and objectives are being met. A summary report for the workshop is available at http://www.nap.edu: *Climate Change Education: Goals, Audiences, and Strategies: A Workshop Summary* (National Research Council, 2011, Washington, DC: The National Academies Press).

The second workshop sponsored by the Roundtable focused on K-14 education, particularly in light of the NRC conceptual framework for new science education standards that included climate change education relevant aspects. A summary report for the workshop is available at http://www.nap.edu: *Climate Change Education in Formal Settings, K-14: A Workshop Summary* (National Research Council, 2012, Washington, DC: The National Academies Press, 2012).

The third workshop focused on how to close the gap in our understanding of climate and sustainability education in postsecondary professional schools of business. A summary report for the workshop is under development and will be available at http://www.nap.edu.

Appendix B

Workshop Agenda

The National Academies
Board on Science Education
Board on Environmental Change and Society
Division of Behavioral and Social Sciences and Education

Engaging Individual, Family, and Community Forestland owners on Issues Related to Climate Change
August 26-27, 2013

Keck Building
500 5th St NW, Room 100
Washington, DC 20001

WORKSHOP AGENDA[1]

Goals:

1. Identify threats from climate change to privately owned forests.
2. Characterize family forestland owners in general, and to the degree possible, in regard to attitudes and dispositions related to forest management and climate change.
3. Discuss the science of two-way communication, adult learning, and engagement that would be relevant for connecting with family forestland owners on the topic of climate change directly or indirectly.
4. Synthesize the previous four goals and identify strategies that extension agents, foresters, and consultants can utilize to successfully engage with family forestland owners on issues related to climate change and forest management.

[1]The agenda will be updated prior to the meeting. Slight changes in timing and order of presentation may occur.

August 26

Introductory Remarks

9:00-9:20 AM	**Welcome to the NRC, Workshop Goals and Objectives**
Speakers:	Martin Storksdieck, Director, Board on Science Education (BOSE) Jim Finley, Committee Chair David Cleaves, U.S. Department of Agriculture

The Changing U.S. Forests: Human and Climate Impacts

During this session speakers will seed a conversation about the changing U.S. forests by providing an overview of (1) the impact of human activity and development on forests and (2) the impact of climate change and natural events on forests.

9:20-10:15 AM	**Overview of Climatic and Human Impacts on U.S. Forests**
Moderator:	Martin Storksdieck, BOSE
Speaker:	James Vose, U.S. Forest Service

Getting to Know Individual, Family, and Community Forestland Owners: Values and Knowledge

Presenters will discuss the values, knowledge, beliefs, management strategies, and other characteristics of individual, family, and community forestland owners.

10:15-11:30 AM	**Data on Forestland Owner Values, Knowledge, and Management Practices**
Moderator:	Shorna Broussard Allred, Committee Member
Speakers:	Brett Butler, University of Massachusetts, Amherst Janean Creighton, Oregon State University Mary Tyrrell, Committee Member, Yale University Geoffrey Feinberg, Yale University
11:30 AM-12:30 PM	**Continued Discussion of Forestland Owners** *(Lunch available)*

Communicating About Forests and Climate Change

Case studies of efforts to get individual, family, and/or community to respond to issues related to climate change will be presented. After the case studies, experts in communication, risk management, behavior change, and relevant fields will discuss how well the case studies align with evidence from their fields.

12:30-3:00 PM **Forestry Communication and Outreach Case Studies: Series 1**

Moderator: Steve Koehn, Committee Member
Speakers: Bill Labich, Highstead
Eli Sagor, University of Minnesota
Tamara Walkingstick, University of Arkansas
Martha Lyman, Consultant (Founding Partner of Community Forest Collaborative)
Allyson Muth, Pennsylvania State University
Don Outen, Baltimore County
Lyle Laverty, The Laverty Group

3:00-4:15 PM **Reflections from Social and Behavioral Science Perspectives**

Moderator: Steve Koehn, Committee Member
Respondents: Maureen McDonough, Committee Member
Purnima Chawla, Center for Nonprofit Strategies

Wrap Up: Day 1

4:15-4:30 PM **Day 1 Wrap Up & Overview of Day 2**

Speakers: Martin Storksdieck, BOSE
Jim Finley, Committee Chair

August 27

Introductory Remarks

8:30-9:00 AM	**Networking and Coffee**
9:00-9:15 AM	**Welcome and Overview of Day 2** Paul Stern, Board on Environmental Change and Society (BECS)

Engaging with Forest Owners

Social and behavioral science experts will present concrete suggestions/proposals on how to address climate change with forestland owners, clarifying audiences, purpose, and actors. Then a panel of diverse stakeholders will respond to the suggestions from the social and behavioral scientists.

9:15-10:00 AM	**Lessons from Social and Behavioral Sciences**
Moderator:	Maureen McDonough, Committee Member
Panelists:	Joe Heimlich, Committee Member Chris Clarke, George Mason University
10:00-11:15 AM	**Reflections from Forestry Stakeholders and Groups**
Respondents:	Claire Layman, Michigan State University Victor Harris, Minority Landowner Magazine James Houser, James Houser Consulting Foresters, LLC Alton Perry, Roanoke Electric Cooperative Amanda Mahaffey, Forest Guild Karl Dalla-Rosa, Forest Service
11:15 AM-12:10 PM	**Break-Out Sessions Focused on Specific Communication Issues**
12:10-1:00 PM	**Continued Discussion of Break-Out Sessions** *(lunch available)*
1:00-1:45 PM	**Synthesis of Break-Out Sessions and Discussion**

Lessons Learned & Next Steps

Important themes and messages and next steps will be discussed. Committee members and the sponsor will provide their input, followed by audience discussion and open plenary remarks.

1:45-2:30 PM **Major Themes and Moving Forward**

Moderator: Paul Stern, BECS
Speakers: Eric Norland, U.S. Department of Agriculture
Jim Finley, Committee Chair
Steve Koehn, Committee Member
Joe Heimlich, Committee Member

2:30-2:45 PM **Small Group Discussions**

2:45-3:30 PM **Open Plenary Remarks**

Moderator: Paul Stern, BECS

3:30-4:00 PM **Final Words**

Speakers: Jim Finley, Committee Chair
Eric Norland, U.S. Department of Agriculture
Martin Storksdieck, BOSE

Appendix C

Registered Workshop Participants

Shorna	Allred	Cornell University
Keith	Argow	National Woodland Owners Association
Theodore	Beauvais	U.S. Forest Service
Patrick	Bixler	Pinchot Institute
Leslie	Boby	Southern Regional Extension Forestry, University of Georgia
Estelle	Bowman	U.S. Forest Service
Brett	Butler	U.S. Forest Service and University of Massachusetts, Amherst
Christine	Cadigan	American Forest Foundation
David	Campbell	National Science Foundation, Division of Research on Learning
Yolanda	Castillo	http://htmlplanet.yolasite.com
Purnima	Chawla	Center for Nonprofit Strategies
Julia V.	Clark	National Science Foundation, Division of Research on Learning
Christopher	Clarke	George Mason University
David	Cleaves	U.S. Forest Service
Kim D.	Coder	Warnell School of Forestry & Natural Resources, University of Georgia
Janean	Creighton	Oregon State University
Karl	Dalla Rosa	U.S. Forest Service
Karen	Dante	U.S. Forest Service

Cathy	Dowd	U.S. Forest Service
Sandy	Farber	University District of Columbia, Cooperative Extension Service
Mary	Farrah	University of the District of Columbia, Cooperative Extension Service
Deborah	Fialka	The American Chestnut Foundation
Geoffrey	Feinberg	Yale University
James	Finley	The Pennsylvania State University
Peter	Gaulke	U.S. Forest Service
Victor	Harris	*Minority Landowner Magazine*, Cierra Publishing Company
Joseph	Heimlich	Ohio State University
Robert	Hershey	Robert L. Hershey, P.E.
Rita	Hite	American Forest Foundation
Colleen	Hoffman	Association of Consulting Foresters
Scot	Holliday	East Meets West Solutions
James	Houser	James Houser Consulting Foresters, LLC
William	Hubbard	University of Georgia, CES - Southern Region
Joshua	Idassi	NC State University, Cooperative Extension
Kenneth	Jolly	Maryland Department of Natural Resources, Forest Service
Michael	Kilgore	University of Minnesota
Steven	Koehn	Maryland Department of Natural Resources, Forest Service
Bill	Labich	Highstead
Lyle	Laverty	The Laverty Group
Claire	Layman	Michigan State University
Monica	Lear	Urban Forestry Administration, District Department of Transportation
Ericka	Luna	U.S. Forest Service
Martha	Lyman	Consultant, Community Forest Collaborative
Amanda	Mahaffey	Forest Guild
Maureen	McDonough	Michigan State University
Andy	McEvoy	Hidden Valley Nature Center
Joanna	Mounce Stancil	U.S. Forest Service
Allyson	Muth	Pennsylvania State University
Frank	Niepold	NOAA, Climate Program Office
Emma	Norland	Cedarloch Research, LLC

Eric	Norland	USDA National Institute of Food and Agriculture, Institute of Bioenergy, Climate and Environment
Don	Outen	Baltimore County Department of Environmental Protection and Sustainability
Alton	Perry	Roanoke Electric Cooperative
Eli	Sagor	University of Minnesota
Al	Sample	Pinchot Institute
Robin	Schoen	National Research Council, Board on Agriculture and Natural Resources
Allison	Silverman	Center for International Environmental Law
Mary	Snieckus	Natural Resource Policy Analysis
Maya	Solomon	U.S. Forest Service
Susan	Stein	U.S. Forest Service
Paul	Stern	National Research Council, Board on Environmental Change and Society
Martin	Storksdieck	National Research Council, Board on Science Education
Anne	Tamalavage	American Geophysical Union
Billy	Thomas	University of Kentucky
Bob	Tjaden	University of Maryland, Environmental Science & Technology
Mary	Tyrrell	Yale University, Global Institute of Sustainable Forestry
Margeau	Valteau	U.S. Forest Service
Allison	Van	Winrock International, Wallace Center
James	Vose	U.S. Forest Service, North Carolina State University
Tamara	Walkingstick	University of Arkansas
Robert	Westover	U.S. Forest Service, Office of Communication
Lynn	Wilson	Association of Consulting Foresters

Appendix D

Workshop Speaker and Steering Committee Member Bios

ARUN AGRAWAL (steering committee member) is a professor in the School of Natural Resources and Environment at the University of Michigan. His research and teaching emphasize the politics of international development, institutional change, and environmental conservation. He coordinates the International Forestry Resources and Institutions network and is a current member of the National Research Council's Board on Environmental Change and Society (BECS).

SHORNA ALLRED (steering committee member) is associate professor of natural resources in the Department of Natural Resources and associate director of the Human Dimensions Research Unit (HDRU) at Cornell University. The goal of her research program is to develop a fundamental understanding of human behavior for the purposes of improving resource conservation and management.

BRETT BUTLER is a research forester with the U.S. Forest Service and an adjunct professor at the University of Massachusetts, Amherst. He is the director of the U.S. Forest Service's National Woodland Owner Survey, co-director of the Family Forest Research Center, and involved in a number of related projects including the Sustaining Family Forests Initiative.

CHARLES CANHAM (steering committee member) is a forest ecologist at the Cary Institute of Ecosystems. His research includes development and application of SORTIE, a spatially explicit model of forest dynamics; neighborhood dynamics of forest ecosystems; effects of vertebrate consumers on forest dynamics; effects of hurricanes on temperate and tropical forest dynamics; watershed-scale analyses of variation in lake chemistry; and likelihood estimation methods and modeling.

PURNIMA CHAWLA is the executive director of the Center for Nonprofit Strategies. Before starting the Center, she served as senior vice president for research and account planning at Equals Three Communications. She currently serves on the board of Citizens for Global Solutions, an advocacy organization.

CHRISTOPHER CLARKE is an assistant professor in the Department of Communication at George Mason University (GMU). He is also affiliated with the GMU Center for Climate Change Communication and specializes in the strategic communication of health and environmental risk.

DAVE CLEAVES is the climate change advisor in the U.S. Forest Service. In this role, he is the primary spokesperson for the agency on climate change and leads the implementation of the Forest Service's nationwide strategy for weaving climate change response into policies, processes, and partnerships. He was formerly the associate deputy chief of Forest Service Research & Development. Prior to his career in the Forest Service, he was a professor of forest marketing and economics at Oregon State University.

JANEAN CREIGHTON is an associate professor and extension specialist in the College of Forestry at Oregon State University (OSU). She is the administrative director for the Northwest Fire Science Consortium and the OSU and USFS PNW partnership. Her research focus includes ways to increase the information delivery and technology transfer capabilities of fuels and fire research to managers and practitioners in the field.

KARL DALLA-ROSA is a forest stewardship program manager with the U.S. Forest Service, responsible for national direction, coordination, and management of the Forest Stewardship Program. Previously, he served as a cooperative resource management forester for Hawaii's Division of Forestry and Wildlife.

GEOFFREY FEINBERG is a research specialist at Yale University's Project on Climate Change Communication (YPCCC). Prior to this, he was a vice president at GfK North America, where he co-founded and was the GfK director of the Associated Press-GfK Poll. In the environmental arena, he served as research consultant to the Sustaining Family Forests Initiative.

JAMES FINLEY (steering committee chair) is Ibberson Professor of Forest Resources at Pennsylvania State University. He served as associate director of the School of Forest Resources, in the College of Agricultural Sciences, and was the extension coordinator from 1997 to 2012. He co-chaired the Roundtable on Sustainable Forests for several years.

VICTOR HARRIS is the publisher and editor of *Minority Landowner* magazine. He has over 20 years' experience in forest management and urban natural resource management, and he was the first African American to work as a forester with the Virginia State Department of Forestry. He also served as the assistant state forester for administration with the North Carolina Forest Service.

JOE E. HEIMLICH (steering committee member) is a professor in extension at Ohio State University (OSU) in the community development unit, an extension specialist at OSU's Center of Science and Industry (COSI), and a professor in the OSU School of Environment and Natural Resources and the Environmental Science Graduate Program. He is a member of the National Research Council's Roundtable on Climate Change Education.

JAMES HOUSER is the owner and consulting forester of James Houser Consulting Foresters, LLC (JHCF), which assists private, nonindustrial landowners in all areas of forest management operations. He serves on the Forest Resource Coordinating Committee.

STEVEN W. KOEHN (steering committee member) is the state forester and director of the Maryland DNR Forest Service. He was president of the National Association of State Foresters from 2009 to 2010 and he currently serves on the boards of directors for the American Forest Foundation and the Sustainable Forestry Initiative, Inc.

BILL LABICH is the regional conservationist at Highstead. He facilitates the development and activity of state-wide and regional conservation partnerships and networks on behalf of the Wildlands and Woodlands Initiative. He has 25 years' experience in the fields of forest management, land use planning, environmental education, and regional land conservation.

LYLE LAVERTY is a consulting forester who provides services to private landowners, agencies, and nonprofits. He is a former U.S. assistant secretary of the interior for fish, wildlife, and parks. He served as the director of Colorado State Parks from 2001 to 2007.

CLAIRE LAYMAN is a public policy education specialist with Michigan State University Extension. She leads its internal climate change outreach team and conducts climate change communication workshops for extension personnel. Recent projects include planning and facilitating community meetings to create climate adaptation plans and conducting listening sessions with farmers on their climate change attitudes and education needs.

MARTHA LYMAN is a consultant in the field of community-based natural resource management. She served as founding partner and coordinator of the Community Forest Collaborative, a partnership of the Trust for Public Land, the Northern Forest Center, and the Quebec-Labrador Foundation/Atlantic Center for the Environment.

AMANDA MAHAFFEY serves as the Northeast region director for the Forest Guild and is a licensed forester in Maine. Most recently, she coordinated a regional conservation partnership of organizations and agencies with a common goal of increasing woodland owners' ability to make informed decisions about their woodlands in Kennebec County, Maine.

MAUREEN H. MCDONOUGH (steering committee member) is a professor and extension specialist in the Department of Forestry at Michigan State University, with adjunct appointments in the Department of Sociology and the Michigan State Museum. Her research and extension interests include community-driven forestry and increasing the diversity of voices in natural resource decision making.

ALLYSON MUTH is a program associate in the forest stewardship program at Pennsylvania State University. In this role she supports an organization of forest stewardship volunteers who promote stewardship to peers. She is also a member of the graduate faculty. Previously she was a decision support forester and assistant district forester for Georgia-Pacific, LLC.

ERIC NORLAND is a national program leader in the Division of Environmental Systems in the USDA National Institute of Food and Agriculture. He also works in the Office of the Climate Change Advisor on a part-time basis as the link to the national extension forestry and natural resources networks and NIFA. He worked for 20 years with Ohio State University Extension.

DON OUTEN is a natural resource manager with the Baltimore County Department of Environmental Protection and Sustainability (EPS) and is the EPS manager for sustainability. He is a member of the Sustainable Forestry Council and participant/core group member of the Roundtable on Sustainable Forests. He is a former deputy director/chief of current planning with the Maryland Department of Planning and Zoning in Harford County, MD.

ALTON PERRY is a forest management-land retention consultant at the Roanoke Center, part of Roanoke Electric Cooperative. He is retired from the North Carolina Forest Service, where he had a number of positions including forest technician, outreach coordinator, and ranger. He worked closely with state and federal partners implementing their forestry/conservation programs and services.

MARY TYRRELL (steering committee member) is the executive director of the Global Institute of Sustainable Forestry at the Yale School of Forestry & Environmental Studies. She leads the Sustaining Family Forests Initiative and serves on the board of the Hamden Land Conservation Trust and on the Scientific and Technical Advisory Committee of the Commonwealth of Massachusetts' Department of Conservation and Recreation, Division of Water Supply Protection.

ELI SAGOR is an extension educator with a focus on forest ecology and management with the Department of Forest Resources at University of Minnesota Extension. He manages the *My Minnesota Woods* website, webinars, and other communications. His work focuses on engaging woodland owners in learning processes that support sound, well-informed land care decisions.

JAMES M. VOSE is a research ecologist and project leader of the USDA Forest Service, Southern Research Station, Center for Integrated Forest Science (CIFS) at North Carolina State University in Raleigh. He spent 25 years at the Coweeta Hydrologic Laboratory studying watershed ecosystem responses to disturbances and forest management. He recently served as co-lead author on the USDA National Climate Assessment Forest Sector Report.

TAMARA WALKINGSTICK is an associate professor of forestry for the University of Arkansas (UA), Division of Agriculture, Cooperative Extension Service (UACES). She is also associate director for the Arkansas Forest Resources Center based at the UA-Monticello, School of Forestry. She serves on the boards of the Arkansas American Indian Center, Arkansas Urban Forestry Council, and National Network of Forest Practitioners.